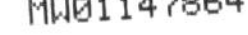

RUDOLF STEINER (1861–1925) called his spiritual philosophy 'anthroposophy', meaning 'wisdom of the human being'. As a highly developed seer, he based his work on direct knowledge and perception of spiritual dimensions. He initiated a modern and universal 'science of spirit', accessible to anyone willing to exercise clear and unprejudiced thinking.

From his spiritual investigations Steiner provided suggestions for the renewal of many activities, including education (both general and special), agriculture, medicine, economics, architecture, science, philosophy, religion and the arts. Today there are thousands of schools, clinics, farms and other organizations involved in practical work based on his principles. His many published works feature his research into the spiritual nature of the human being, the evolution of the world and humanity, and methods of personal development. Steiner wrote some 30 books and delivered over 6000 lectures across Europe. In 1924 he founded the General Anthroposophical Society, which today has branches throughout the world.

THE WORLD OF BEES

RUDOLF STEINER

Compiled, with commentaries, by Martin Dettli

Translated by Matthew Barton

RUDOLF STEINER PRESS

Rudolf Steiner Press,
Hillside House, The Square
Forest Row, RH18 5ES

www.rudolfsteinerpress.com

Published by Rudolf Steiner Press 2017

Originally published in German under the title *Die Welt der Bienen* by
Rudolf Steiner Verlag, Basel, in 2010

A catalogue record for this book is available from the British Library

Print book ISBN: 978 1 85584 540 4
Ebook ISBN: 978 1 85584 498 8

Cover by Morgan Creative featuring a photograph by
Muhammad Mahdi Karim
Typeset by DP Photosetting, Neath, West Glamorgan
Printed and bound by 4Edge Ltd., Essex

CONTENTS

INTRODUCTION

From time immemorial, people have been fascinated by bees. Mythic pictures and writings tell of our close affinity and connection with these heavenly beings and the inestimable value of honey and wax. And these myths still retain their potency today, even if we now have different explanations of the life of bees, and their coexistence with us.

In recent years these creatures have come to prominence again in the media: reports of colony collapse and the demise of bees have been dramatized through the quote attributed to Einstein: 'If bees disappear,' he is supposed to have said, 'humankind will only survive for four more years; without bees, no plants, no animals, and no human beings.' This quote had a dramatic effect: the mix of prophecy, science and myth shifted bees into the forefront of the popular imagination. Bees now figure in politics, science and culture. And they need this attention, for they are ailing, as we can see from frequent reports of colony collapse.

Beekeepers in particular, in their interactions with the 'bee-ing' or the bee 'super-organism', understand the special relationship between humans and bees. There is ritual character even in their preparations to meet this being, such as putting on white clothes, lighting the smoke canister, and the moment of turning to the hive. Cultivating a calm and peaceful mood, beekeepers immerse themselves in the aura and atmosphere of the bee-folk, experiencing the world of fragrances that issue from the hive, and its peaceful hum. How one moves is important too, since a calm and confident approach elicits a similar response from the bees, reducing the danger of being stung. Meeting the world of the bees we enter a different dimension, often losing our ordinary sense of time. All the more painful then is the rude awakening when we find that bees are ailing through loss of wild flowers, diseases or parasites.

The strong connection between people and bees goes still further though. The hive is an astonishing community at many levels. Simply the fact that bees live from and through flowers can touch us: creatures whose existence draws on this wealth of colour and fragrance without ever harming it. Quite the opposite, for through pollination they ensure that blossoms bear fruit. Honey bees, bumble bees and butterflies are the only beings that find nourishment without needing to destroy or rob other living creatures.

Then there are the amazing parallels between developmental stages in bees and vertebrates including the human being. Out of themselves they create their comb as a kind of sustaining organ. The cells are built from their own body substance, and we can see a correspondence here with the bone structure of vertebrates. The bees can regulate their body warmth and maintain a brood warmth of 36 degrees, like warm-blooded creatures. The way that the brood are nourished by gland secretions of the nursing bees resembles mammalian life. In biology these three parallels are seen as the sign of increasing independence from direct environmental influences. In this context it is interesting that, like humankind, beekind has succeeded in adapting to almost every climatic zone.

The bee community's organization shows various parallels with human society and yet is also very different, and this in itself makes our study of it so valuable. The differences can stimulate ideas, impulses and new approaches—in medicine, food processing, the communications field, and in the realm of decision-making. The bee-folk provide inspiration for medicines development, technical solutions and social innovation, and offer us allegories that make the working of the world of spirit a tangible reality to us—for instance a departing swam as a metaphor for the soul leaving the body at death.

The bees form an organism. As a whole being they create an organ in the comb that supports and sustains them, that acts both as storage facility and cradle for the brood. Other 'organ functions' are accomplished by the common endeavours of the worker bees. At different ages and stages in their life, worker bees take on differing tasks in the service of the whole hive; they begin their work inside the hive and end it as foragers out in the wide world. Their first tasks involve cleaning comb cells close to where they have recently hatched. After this they nurse the young brood and develop feeding glands for their wet-nurse duties. Later these glands cease functioning and instead their wax glands are activated: the wet-nurses turn into builder bees. The transition towards the outside world occurs when they take up posts as guard bees at the hive entrance; and only when they reach a certain age do they take wing into the surrounding landscape to visit flowers and bring back to the hive the nectar, pollen and water so vital for its survival. In the summer months bees live scarcely more than a month, so that all the hive members apart from the queen are completely renewed in a very short time. As they rear their young, they pass on their life to the new generation. When brood-rearing declines, the bees live for longer: between July and September 'winter bees' hatch who can live for as long as seven months.

The female workers are not responsible for all the hive's organic functions. They are kept apart from sexual life, which is instead the task of specialized bees. The only female with reproductive capacity in the hive is the queen, while the drones take care of the male side of the contract. The fate of the two sexes is very different: the queen, as organ of unity, is of central importance, and lives in the hive for several years. The drones, on the other hand, are only present in larger numbers for about two months in early spring.

The complex organization of inner life processes in the bee-ing cannot be assigned to any one governing organ, and the same is true of the behaviour of the hive in general. Responses to their surroundings and necessary 'decisions' occur at a communicative level that is not centrally dictated.

Hives have a wide range of possible behaviours: depending on the situation, the community of one hive may react in one way, while another reacts differently, revealing recognizable individual characteristics.

Our wonder about bees can lead to many questions, and in the search for greater understanding, which can help us develop a different way of relating to them, we find much of great value in Rudolf Steiner's insights. In his works there are diverse references to bees and the life of the hive, often drawing on imaginative metaphors. His lectures to workers at the Goetheanum gave rise to a series entirely devoted to bees. Responding to a concern from beekeepers in his team of builders, he took their questions and descriptions as his point of departure, and the content, structure and wording of these lectures is therefore different from written works. Since the talks were taken down in shorthand, errors may occasionally have crept in to the texts published here. But the sequence of lectures testifies to a structure that we have retained: they are presented in their entirety and form the main framework for this book. In between, some of the themes that surface in the lectures are taken up with commentaries drawn from present perspectives, augmented by further passages from Steiner's works. The fascination of these lectures on the nature of bees lies not least in their multi-layered quality, including ideas that are enigmatic and hard to categorize but which can spur us on to deeper enquiries, as well as intellectually more accessible ones. Others in turn can be taken up with full warmth of heart, and can leave a deep impression on us. From the standpoint of beekeepers it is also lovely to discover some useful practical suggestions for working with hives.

I hope that, as a kind of 'handbook', this compilation can awaken many people's interest and prove useful to them. Practical guidance about natural beekeeping and creating a new colony from a swarm can be found on my website at www.summ-summ.ch

I would especially like to thank Johannes Wirz for his helpful suggestions and Xaver Wirth, now deceased, who was my companion for many years in the quest to understand the nature of the bee-ing. It was with him that I first started this project.

Martin Dettli

1. THE ORGANISM OF THE HIVE

The social life of bees and their wise collaboration is fascinating to observe. Bees are in harmony with each other: they live in intense contact and vibrant communication. The life of the hive depends on fine-tuned organization. The colony lives in darkness, in some hidden corner, in a tree trunk or a beehive.

The bees fly out one by one into the light, seek flowers or fetch water. When they swarm, on the other hand, they appear together in the light in a whole, expansive swathe, whirring through the air and eventually gathering again in a cluster.

What exactly is this phenomenon of a colony?
Does it consist of many individual bees that coordinate socially with each other, or is it really a whole organism, of which each bee is an integral part? It is not easy to answer this. As a beekeeper one easily moves back and forth between the two possibilities.

When we open the beehive, we see that the colony is constituted of distinct and separate bees. We see the workers primarily, but may also glimpse the queen and some drones. The bees live on combs in which they nurse their brood and store their supplies. We can remove the combs and dismember the unity into parts. If we start from the separate creatures, and bring them all together in our thoughts as a unified whole, we arrive at the principle of self-organization.

Modern science studies the colony in these terms.[1] According to this view, the many single bees create a new whole through their collaboration and communication. This cooperative activity forms what we call the colony, which possesses qualities far exceeding those of each separate creature and which, as a whole, works back in turn on its parts. This idea of self-organization contrasts with the approach taken by Rudolf Steiner.

The colony is the whole; the workers, queen and drones are parts that serve it, rather as our cells are part of us. The unity is superordinate. The holistic view of what occurs in a colony can be confirmed most clearly in the swarm cluster: there it hangs, rounded in shape, resembling a pear. In the close coherence of all the many bees the colony shows its will to form a new unity. If you catch this swarm and lodge it in a hive, and allow it to build its own combs, these will grow like whitish internal 'organs'. At this moment we can have a moving sense of its single unity. Later on we can again approach this unified bee-ing by eavesdropping on the hive or observing what happens at its entrance. A strong sense of this unity can be gained especially from a colony hidden within its enclosed structure.

If we consider ourselves and other vertebrates, we can imagine that our life processes and soul stirrings are governed from somewhere within the body, while our sense organs connect this inner world with our outer surroundings. When we make contact with an animal of some kind, we can observe its eyes, body language and behaviour. This is not so self-evident in the bee colony. The place where coordination of the hive's body and soul occurs cannot be physically located. While we experience its unity we cannot grasp it physically. It is not to be found in the separate bee, and so it must lie somewhere between all the creatures. This compels us to imagine a spiritual bond that unifies them—which is what ultimately characterizes the other way of looking at bees.

Rudolf Steiner always pointed to this organism. Beekeepers do not necessarily find it self-evident to see the

> colony as a single whole, and this represents a challenge to them in their daily work. To cultivate a sense of this unity, we have to keep focusing on it. But if we do, this continual view of the hive as a whole organism has practical consequences: it requires a way of handling bees adapted to the life of an organism. Thus we focus primarily on the life of the whole 'individual' colony and its impulses.

So here it is indeed important to consider something such as the hive and learn that the single bee is in itself dull-witted. It has instincts, but is dull-witted; yet the whole hive is extraordinarily wise. You see, recently we had some very interesting discussions amongst the workers, to whom I regularly give two talks each week. We were discussing bees, and a very interesting question came up, one that a beekeeper knows is of great importance. If a beekeeper beloved by the bees falls ill or dies, the whole hive comes into disarray. Now there was someone at the talk with a very modern kind of outlook, and he said this: 'But the bee can't really see clearly, it has no concept of the beekeeper, so how could it have any sense of belonging? And moreover, the colony a beekeeper is tending will be a quite different one each year: apart from the queen, none of the bees will be the same, so how can any sense of belonging develop?' I replied as follows: If you understand the human organism you will know that it renews all its substances at regular periods. Imagine that you get to know someone who sets sail for America and returns ten years later. You will have before you a quite different person from the one you knew ten years ago. All the substances that compose him have changed, and he presents you with a quite new configuration. This is nothing other than what we find with the hive, in which the bees are now different ones.

Nevertheless a sense of belonging survives between the bees and the beekeeper. This sense of belonging is based on the great wisdom living in the hive: it is not just a little heap of single bees, but the hive really does possess a distinct, tangible soul.

This is something we need to incorporate again into our sense of nature: this view that the hive has a soul.[2]

But it is also true to say that the hive lives its own, very remarkable life. What is this due to?

You cannot begin to explain this unless you can perceive the spiritual realm. Life in a beehive is organized with extraordinary wisdom—and anyone who observes the life of bees will concur with this. Of course we can't say that bees have a science or body of knowledge like we do, since they do not possess a brain like ours and everything connected with it. They cannot drink in universal reason as we do and embody it. Yet influences from their whole environment have a hugely powerful effect on the hive. If we understood that everything in the environment exerts a huge effect on conditions as they exist in the hive, we would get a proper sense of the nature of the life of bees. Far more than ants or wasps, the life of the hive depends on the bees really collaborating very closely, accomplishing their work so that everything is in harmony. And if we seek to discover why this is, we find that what other animals express in sexual life is suppressed to a very great degree in the life of bees, suppressed to an extraordinary extent.

Reproduction, you see, is taken care of by only a very few, specially selected individuals, the queens. In the others, sexual life is more or less suppressed. But in sexual life, love exists as something initially inward. Only when this soul element pervades certain bodily organs do they come to manifest or express love. Because the bees suppress their love life, with the sole exception of the queen, sexual life in the hive is transformed into all this busy activity which the bees develop together. This is why people of more ancient and wiser times, who knew things in a quite different way from how we know them today, why these wise ancients assigned the whole wonderful activity of the hive to love, to the life that they connected with the planet Venus.

And so we can say that wasps and ants tend to withdraw themselves from the influence of the planet Venus, whereas bees are wholly given up to it, and develop it throughout the hive. This life is full of wisdom—you can imagine yourself what wisdom lives there. In my various accounts of how a new generation arises, I have shown the unconscious wisdom living in this. The bees develop this unconscious wisdom in their outward activity. What only unfolds and is embodied in our heart when we develop love becomes something like a substance in the whole hive. The whole hive is really pervaded by the life of love. The individual bees relinquish love but develop it instead throughout the hive. And so we start to understand bee existence if we recognize that the bee lives in an air, an atmosphere, that is entirely impregnated with love.

Moreover the bee benefits most of all through living, really, from the parts of the plant that are again, in turn, entirely pervaded by love. The bee sucks the nourishment which it then turns into honey from the parts of the plant that serve love, and thus as it were brings the love life of flowers into the hive.

And so the life of bees has to be studied soulfully.

This is far less so when we study the ants and the wasps. We will discover that these creatures, by contrast, withdraw from what I described in the case of bees, and instead give themselves up more to sexual life. Apart from the queen bee, bees are really creatures that would say, if they could speak: We wish to relinquish our own sexual life and make ourselves into the bearers of love.

They bear into the hive what lives in the flowers. And if you start to think this through properly, you uncover the whole secret of the hive: the life of this burgeoning, springing love that is dispersed in flowers is then also contained in honey.[3]

2. BEEKEEPING

In central Europe, bees were originally kept either in round baskets called skeps or in a piece of hollow tree trunk, the beegum. Both were a simple, single unit. In those days, an important part of the beekeeper's work was to catch swarms and rehouse them. The colony would build its combs so that they adhered directly to the underside of the skep, as a fixed comb. To harvest the honey, the basket was turned upside down and part of the comb was cut out. Often this was only done after the colony had overwintered, once its survival was assured. Skeps were usually set up along the outer wall of a dwelling, and human beings and bees lived in close proximity.

In the middle of the nineteenth century, a new period dawned for beekeeping with the invention of moveable frames so that combs could be taken out and replaced. This also made it easier to observe the life of the colony, and knowledge about bee biology increased. In the second half of the nineteenth century, the moveable frame was followed in quick succession by wax 'comb foundation' sheets with the hexagonal cell pattern pre-printed on them, then the honey separator or extractor, and artificial breeding of queens. All these inventions continue to be used today, and there were few new developments in the twentieth century.

The moveable frame gave rise to a different kind of beekeeping. The rectangular frame is placed in a box or a chest and can be moved and taken out at will. The invention of the honey extractor meant honey could be harvested free of wax, and without destroying the combs. A range of procedures became possible. Storage combs filled with honey can be moved from one colony to a different one to strengthen other hives, or to propagate a new one. And a whole colony can be divided into several new ones.

The comb foundation is a wax sheet intended to make it easier for the bees to build their combs since they need to produce less wax themselves. And yet at the same time they are compelled to organize their work precisely in line with the predetermined cell dimensions on the sheet. Drone breeding, which requires the building of larger cells, is reduced to a single comb so that their numbers are kept in careful check.

Artificial queen breeding was invented in America at the end of the nineteenth century to meet a great demand for Italian bees in the new continent. For this purpose normal larvae from the worker-bee brood are moved at an early stage into artificially created queen cells, and placed in a colony from which the queen has been removed or that was created without one. In this way colonies can be compelled to raise new queens. This makes it easier to multiply selected bee strains. Once hatched, the queens are infiltrated into small, artificial colony units for procreation, and are a desirable commodity. They can form a new colony with arbitrary other bees, or be introduced into a hive as a replacement queen.

All these procedures are standard in modern beekeeping. Instead of an outlook oriented to a living organism, a strongly mechanistic approach prevails and accordingly determines how people relate to their bees. Everything can be moved around or exchanged, the whole can be divided into parts and re-amalgamated. This applies to everything—honeycombs or brood combs, worker bees or queens. The analytic outlook leads to beekeeping that is perhaps reminiscent of factory work: functionality is key. Rather than caring for and cultivating the organism, the focus is on the colony's productivity.

The following comments by Rudolf Steiner came at the end of a talk on the state of modern beekeeping given by a beekeeper named Müller on 10 November 1923.

The thing is this—and I'll speak more about it next time: honey production, the whole work connected with it and even the industry of the worker bees can be enormously increased by artificial apiculture methods. But, as Mr Müller has already remarked, it should not be undertaken in too strictly rational a way nor made overly commercial. Next time we will take a closer look at beekeeping and will see that what can be an extremely beneficial measure for a short period will seem a good idea today, yet in a hundred years all beekeeping would come to an end if only artificially bred bees were used. It is worth examining how something that is extremely beneficial in the short term gradually, over time, comes to be something that kills off the whole business. At the same time let us consider how beekeeping is of especial interest in helping us understand all nature's secrets, in particular how something that appears to be enormously fruitful on the one hand can actually lead on the other to its demise.

So while beekeepers can take great delight in the advances introduced into apiculture in recent times, this delight will not last for even a hundred years.[4]

I wanted to make a few remarks relating to Mr Müller's talk. I think they may be of interest even though, of course, there is insufficient time today to really apply such things in practical apiculture. I have very little more to add—in fact nothing at all really—on practical aspects of beekeeping, since Mr Müller already gave a very fine account of modern practices.

But if you were listening carefully, you will have been struck by the whole enigmatic world and nature of beekeeping. The beekeeper of course is primarily interested in the work he must do. But all of us should have the greatest interest in beekeeping since in truth more depends on it for human life than we usually realize. Let us take a step back for a moment and consider the matter in broader terms. As you heard from what Mr Müller told you, the bees can gather what is already contained, really, as honey in flowers.[5] They collect this nectar, and then we humans take from them a part, only, of what they harvest in the hive, not even a very large part of it. We can say that beekeepers remove roughly 20 per cent. But then the bees are also furnished with a physical organization that enables them to take pollen from the flowers. Thus bees gather from flowers something that these contain only in small amounts, which is very hard to harvest. The relatively tiny quantities of pollen are gathered by bees in the pollen sacs they have on their rear legs, and are stored and then consumed within the hive. In the bee, therefore, we find a creature who collects a substance present in nature in very fine and dispersed form, and then makes its own use of it in the hive.

But let us continue. After the bee—something people scarcely notice perhaps because they do not consider it—has transformed its food within its own digestive tract into wax (for bees produce wax themselves), it creates its own small cell or vessel where either eggs are laid or supplies are stored. This little vessel is, I would say, very remarkable. From above it appears hexagonal, and seen from the side it looks like this [see drawing]. It is sealed at one end, and eggs can be laid there or supplies placed inside it, the two purposes side by side. These things go very well together so that in the comb extremely good use of space is made by these adjoining cells.

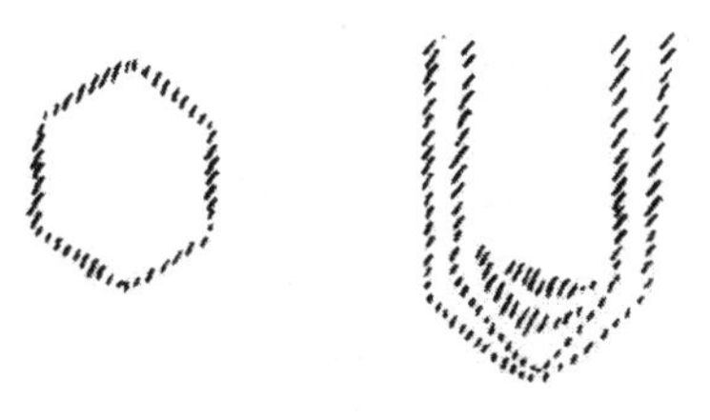

But if we ask how the bee can instinctively build a cell constructed in this way, people usually just reply that it is in order to make best use of space. And that is true too. If you tried to think of a different cell shape, there would always be some space intervening between adjoining cells. Here no superfluous space is left; instead everything adjoins to make full use of space in the comb.

Yet that is not the only reason. Imagine the little grub, the larva completely enclosed in the cell; you should not think that what exists in nature exerts no forces. This whole hexagonal case, this six-sided casing, possesses forces. Things would be quite different if the larva were lying in a sphere. In nature it has a quite different significance for the larva to be contained in these hexagonal forms, which enter and inform it. After hatching, the bee will sense in her body that in her infancy, when she was softest, she dwelt in this hexagonal cell. And with the same power she imbibes there she then builds other such cells. They contain the very forces through which the bee works, and the environment that the bee outwardly creates. Well, that is the first thing we should consider.

Then you heard another, very remarkable fact: there are different kinds of cells in the hive. I am sure that a beekeeper can easily distinguish worker cells from drone cells. And it is still easier to tell the cells of workers and drones apart from queen cells, since the queen cells have a quite different shape, more like a sack. There are very few of these in a hive. So we find that while the workers and drones—the drones are the males—develop in six-sided cells, the queen grows really inside a sack. She is as it were exempt from the otherwise angular planes of her environment.

Moreover, to fully develop, to become fully-fledged, the queen takes only 16 days, while a worker bee needs longer—21 days. It looks like nature takes much more care, therefore, in developing worker bees than the queen. We will see in a moment that there is another reason for this. The workers need 21 days to develop. And the drones, the male bees, whose lifespan is shortest, who are killed off after they have fulfilled their task, actually need between 23 and 24 days to develop.

This is worth noting: the different types of bee, queen, workers, drones, require a different number of days for their development.

Now these 21 days required for maturation of the worker bees have a very particular significance: 21 days is not a random number of days for something occurring on the earth—in fact it is the period during which the sun rotates once on its own axis.

The worker bee develops, therefore, in exactly the same time as the sun takes to rotate once. The worker bee participates in a whole revolution of the sun and in consequence enters into everything that the sun can effect in her. Picture the worker bee: when its egg is laid, its point relative to the sun at that moment recurs again after 21 days. If it were to continue in this influence, what issues from the sun would be a recurrence of what has already been. And so, for its full development, the worker bee enjoys everything the sun can accomplish. If the worker were to go on pupating after this period, she would leave the sun's influence and enter into an earthly developmental phase, no longer engaging in sun development since she has already had that, already fully savoured it. In fact she experiences this only as a fully-fledged insect, a mature creature now. You might say that she takes a tiny earth moment for herself and then, after this solar development, becomes entirely a creature of the sun.

But now consider the drone, who, if you like, reflects a little further on things. He decides that he is not ready after 21 days. Before completing pupation he enters therefore into earthly development, so becoming a creature of the earth. The worker bee is a completed sun creature.

And what about the queen? She does not even complete the whole cycle of sun development. She remains behind, always remains a sun creature. In other words the queen always remains closer to her larval state, her grub state, than the other bees, while the drone is furthest removed from grubhood. By remaining closer to her larval state the queen becomes able to lay her eggs. So you can see clearly in bee life what it means to be subject to solar or earth influence, for whether a bee becomes a queen, a worker or a drone depends simply on whether she awaits a full cycle of sun development or does not do so. Because sun development remains present to her always, and she assimilates nothing of earth development, the queen can lay eggs. The worker bee develops a little longer, four or five days more. She savours the sun entirely. But at the moment her body is solid enough, she does touch into earth development momentarily. Her full absorption means she cannot return to sun development and so cannot lay eggs.

The drones are the little males, and can fertilize. This fertilization is earth-related therefore. The drones gain their powers of fertilization through the couple of days longer that they pupate in an earth development phase, not once they are mature. So by studying bees we see clearly that fertilization by the male comes from the powers of the earth, whereas the female capacity to lay eggs comes from solar forces.

And so, you see, in this way you can gauge the significance of the length of time a creature takes to develop. This is very important since something occurs in a certain period that does not occur over a shorter or longer period; something else happens.

But there is more than this. The queen, as we see, develops in 16 days. We don't get to the point in the sun cycle she started from until after she hatches, so she remains within sun development. The worker bees complete the cycle but they stay within sun influence, rather than continuing on into earth development. Thus they feel an affinity with the queen. Because they belong to the same sun development as she does, the whole hive of workers feels an affinity with the queen, feels connected with her. The drones, they would say, are turncoats, they have already fallen earthwards and no longer really belong to us. We only tolerate them because we need them. And why do they need them?

Occasionally it happens that a queen is not fertilized but she still lays eggs that can develop. The queen does not have to be fertilized, she still lays eggs. This happens to some extent with other insects too, and beekeepers call this a virgin brood since the queen is not fertilized. The scientific name for this is parthenogenesis. But only drones hatch from unfertilized eggs, not workers or queens. So if the queen is not fertilized no further workers or queens can emerge, only drones, and a hive like that would naturally be useless.

A virgin brood always produces only the other sex and not

the same sex. That is a very interesting fact; in the whole balance of nature, fertilization is vital for producing the other sex—in lower animals of course, not higher ones. Only drones hatch from bee eggs that are not fertilized.

Fertilization itself is something very special amongst the bees. Rather than some kind of marriage bed from which the guests retire while fertilization is accomplished, it occurs in full view, in the brightest sunshine and, strange to relate, as high up in the sky as possible. The queen flies as high as possible towards the sun to which she belongs. I have described this to you. And the drones that can overcome their earth forces—for it is with these that they have united—and fly the highest, can fertilize the queen high up in the sky. Then the queen returns and lays her eggs. So the bees don't have a marriage bed but a nuptial flight, and when they wish to perform fertilization they try to fly sunwards. Good weather is also needed for this—the sun really is needed—and this event does not happen in bad weather.

All this can show us how sun-related the queen remains. After fertilization has happened as I described, worker bees are produced in the corresponding worker brood cells. First, as Mr Müller described to you very well, the little grubs form, and so on, and in 21 days they develop into worker bees. In the sacklike cells a queen develops.

Now to take this further I will have to say something that will of course seem a little dubious to you at first, since further study is needed to confirm it. Yet this is true. Once the worker bee has grown to maturity and flies out to forage, she uses her tarsus or foot claws to gain purchase and then she can imbibe the nectar and collect pollen. She places the pollen in the pollen sacs on her rear legs, but she sucks up honey with her proboscis. Some of it serves for her own food but most of it is stored in her honey stomach, and she regurgitates it when she returns to the hive. When we eat honey, actually, it is worth noting that we are eating bee sputum—but a very pure, sweet sputum, unlike other kinds! So the bee gathers what is needed for food or for storage, for processing, and for wax production and so forth.

But now we have to ask how the bee finds her way to a flower. She makes a beeline for it, unerringly. This cannot be explained just by studying the bees' eyes. The bee, I mean the worker—drones have somewhat larger eyes—has a small eye on each side, and three very tiny ones on the top of its head. If you study these eyes of the worker bee you find they cannot see very much, and the three small, tiny eyes [ocelli] cannot see anything at all. So the remarkable thing here is that the bee does not find her way to the flower by eyesight but through something resembling smell. She feels her way towards fragrance and finds the flower. A sense somewhere between smell and taste leads the bee to a flower from a good distance away; and this is why the bee does not need to use its eyes for this.

Now picture this as clearly as you can. A queen has hatched, born in the sun-realm, having not fully savoured the sun influence but as it were remaining in it. Then a whole army of workers developed further under the sun's influence but without passing over into earth development. These workers feel connected with the queen—not just because they were subject to the same sun but because they remained within sun development itself. In their development they did not separate from the queen's development. The drones do not belong to this—they went further and separated themselves from this realm.

And now the following happens. When a new queen

emerges, the nuptial flight must take place. The queen emerges into the sunshine, a new queen is born. For the whole host of worker bees who feel themselves connected with the old queen, something very curious happens: their tiny eyes become seeing when a new queen emerges. The bees cannot endure this: they cannot bear for something the same as themselves to come from somewhere else. The three tiny eyes on their head, these three, really tiny eyes, form entirely from within the worker bees and are pervaded by the inner bee blood and so forth. They have not been exposed to the external influence of the sun. These tiny eyes of the bees suddenly become what one might call clairvoyant by virtue of the fact that the new queen, born from the sun, brings sunlight into the hive with her own body; the bees cannot endure this light from the new queen. And so now a whole swarm forms—it is something like fear of the new queen, as if they were blinded by her light, as when we look directly at the sun. And therefore they swarm off; and then one has to establish a hive again with the old queen, with the social cohesion of most of the worker bees who still belong with the old queen. The new queen has to cultivate a new colony for herself.

Some of the bees remain behind [with the new queen] but they are the ones who hatched under different conditions. Bees basically swarm because they cannot endure the new queen who introduces a new sun influence into the hive.

And if you ask why the bees are so sensitive to this new sun influence, you discover something very remarkable. It can of course sometimes be a little uncomfortable to make a bee's acquaintance, to be stung. A large creature like the human being only suffers a red sore on the skin, unpleasant though this is. Small creatures can die from it, and this is because the bee has a stinger which is really a tube, within which something like a plunger rises and falls, conducting poison from the bee's poison sac.

This poison, unpleasant to creatures that suffer it, is of extraordinary importance for the bees. Actually they don't particularly enjoy dispensing the poison when they sting, but they do so because they cannot endure any outer influence. They wish to remain within themselves, in the world of their hive, and they experience every outer influence as a disturbance which they fend off with their poison. This poison however continually has another significance too: it disperses in tiny amounts always through the bee's whole body, and without it she could not survive at all. If you study a worker bee, you have to recognize that she cannot see with her tiny eyes, and this is because the poison continually enters these ocelli. The action of the poison is impaired at the moment when the new queen arrives with the new sun influence—it ceases to work and these eyes therefore suddenly gain the power of vision. And therefore the bee really owes to its poison the fact that it lives continually in a kind of twilight.

To give you a picture of what the bees experience when a new queen hatches from a sacklike queen cell, I would have to put it like this. A bee lives always in twilight, feeling its way with its sense of taste-smell (something halfway between smell and taste), feeling its way around in a kind of twilight which it feels at home in. But then the new queen arrives, and this is rather like going out in the dark in July and seeing luminous fireflies or glow-worms shining. That's how the new queen shines to the bee colony, since the poison in them no longer acts strongly enough to sustain them inwardly. The bee needs seclusion from the world, twilight seclusion, and has this even when she flies out of the hive since her poison keeps her self-possessed. And then she needs this poison when afraid that something is approaching her from without. The colony wishes to be entirely self-enclosed.

For the queen to remain in the sun realm, she must not be in a hexagonal cell, either, but in a rounded cell, where she remains entirely subject to the sun's influence.

And now we come to something that must make beekeeping of great interest to everyone. What happens in the hive, you see, closely resembles what happens in our own head, albeit with a slight difference. In our head, the substances don't form the same outgrowths. But we have, don't we, nerves, blood vessels and then also isolated protein cells, which retain a rounded shape—they too are always there somewhere. We too have three things inside our head. The nerves likewise consist of single cells which do not grow fully into creatures since nature restrains them—but these nerves would become creatures if they could. And if the nerve cells of the human head were able to develop in all directions under the same conditions as exist in the hive, these nerve cells would become drones. The blood cells flowing in our veins would become worker bees, and the protein cells, present especially in the midbrain, pass through the shortest development and can be compared with the queen; and thus we have these same three forces at work in the human head.

Now the worker bees bring what they gather from flowers back to the hive, process it in their own body into wax, and create this whole wonderful comb. This is also what the blood cells in the human head do! They pass from the head into our whole body; and if you take a look at a piece of bone, for instance, you can see the hexagonal cells in it. The blood circulating in our body undertakes the same work as bees in the hive, except that in other cells, in the muscles where conditions are similar—since muscle cells also resemble the wax cells of the bees—they dissolve too soon, remaining soft, and so we don't discern this in the same way. In the bones we can detect it certainly. And so our blood has the same forces that a worker bee possesses.

Yes, you can even study this in connection with time. Cells present in the earliest stages of embryonic development, and surviving into later stages, are protein cells. The others, the blood cells, arise somewhat later; and last of all the nerve cells develop—and this is just what occurs in the hive! The only difference is that the human being builds up a body that seemingly belongs to him, while in the bee this body is composed of the cells of the comb. With this comb structure the same occurs as inside our body, except that there it is not so easy to demonstrate that the blood cells do this out of a kind of wax. We ourselves are made out of a kind of wax, as bees form it in their skep or box. So we can put it like this. The human being has a head, and the head works upon the larger body, which is really the hive. In the hive, the connection between the queen and the workers has the same relationship which protein cells—which remain round—have with the blood in us. And the nerves are continually being ruined, worn out, since we wear out our nervous system. It is not exactly that we wage a nerve battle within us as the bees wage a battle on the drones—for then we would die every year. But nevertheless our nerves grow weaker each year. And we die, really, from our weakening nerves—we can no longer feel our body in the same way, and die from wearing out our nerves.

If you consider the head, which represents the hive really, you find that everything in it is enclosed and protected. And if you introduce something into it from outside, this is a dreadful injury to it. The head won't endure it. This process that happens when the new queen emerges is something the hive can't endure either—the colony would rather take flight than live with this new queen.

It is for this very reason that beekeeping should be regarded as something enormously important. As we have heard, beekeepers take about 20 per cent of the honey the bees produce, and this honey is extraordinarily useful to us. Without it we would obtain very little nectar in our food, since it exists only in tiny quantities in plants. We also have 'bees' inside us, in our blood; and this bears the honey we eat to the diverse parts of the body. But this honey is what the bees need to make wax so that they can build the 'body' of the hive, its comb structure.

Especially as we grow old—in the child, milk has the same effect—honey is extremely beneficial for us, nurturing our bodily constitution. Honey is highly recommended for old people, though not in excess. If one eats too much of it instead of using it just as a condiment, this results in too much formative action: our form grows brittle and we may get all sorts of illnesses. A healthy person will sense how much he can eat; and honey is a very healthy food for the elderly since it lends the body stability, real stability and solidity.

Children with rickets would benefit from the following—not in earliest infancy, when children ought really only to live on milk, and honey won't as yet take effect, but at the age of nine or ten months. At that age, if one were to give them honey in carefully measured amounts, and continue with a honey diet until the age of three or four, rickets would not assume such severe proportions. This is because rickets indicates a body that remains too soft and collapses in itself. Honey contains the power to form the human body, to give it stability. We really should gain insight into such things and, if we did so, much, much more attention would be given to beekeeping than currently is the case.

The following is also possible. In nature, you see, everything is interconnected in a remarkable way, and the laws which human beings cannot penetrate with their ordinary reasoning are actually the most important of all. It is true to say that these laws always leave a tiny scope for freedom. This is true, for instance, of the genders on earth. There is not exactly the same number of men and women on earth, but approximately equal numbers across the globe. Nature's wisdom itself ensures this. If things were to come to a point—I think I have mentioned this to you before—where human beings themselves could determine the sex of children born, things would quickly come into disarray. Nature itself regulates these things—so that, for instance, in regions where wars have decimated the population, fertility levels rise. In nature every lack or deficiency elicits the power to remedy the imbalance.

When bees gather honey in a particular region, naturally they take nectar from the plants they visit—but take it from plants otherwise useful to us, fruiting plants and suchlike. In regions where bees are kept, fruit trees and similar plants thrive better than in other places. When bees take honey from plants, nature does not sit back in leisurely fashion but creates more such fruit-bearing plants, and so we not only benefit from the honey which bees give us, but gain more from the plants visited by them. This is an important law we can easily understand. And when we do, we can recognize that the interrelationship between bees and plants, this reciprocal organism, contains wonderful wisdom. Bees are subject to natural forces that are extremely important and really wonderful; and when we recognize this we feel a certain reluctance to intervene in these natural forces.

Wherever we intervene in natural forces we do not make things better but worse. We do not immediately make things worse, but everywhere nature keeps working as well as she can despite the hindrances she encounters. We human beings can remove certain hindrances, and make things easier for nature in certain respects. I am thinking here, for instance, of modern bee boxes which have replaced the old bee skeps and apparently make things much easier for beekeeping.

But let us consider artificial bee breeding. Don't think that I can't see the initial advantages of this, in practical terms quite irrespective of spiritual science: naturally it makes some things easier. But in the long term it will impair the strong cohesion of a bee colony. Today of course, in some respects, there can be nothing but praise for artificial bee breeding as long as all the precautions Mr Müller referred to are taken. But we must wait to see how things will be in 50 or 80 years' time; for certain forces, you see, that previously worked organically in the hive are being mechanized, are being done mechanically. When a queen is purchased and introduced into the hive, the colony can no longer create the same inner affinity between queen and workers as it did when the colony itself produced a queen. But to begin with there will be no noticeable effects.

Of course I have no desire whatever to launch a fanatic movement opposed to artificial bee breeding. Such a thing would be impracticable—roughly like the following. You can more or less work out when the earth's coal resources will be exhausted; they will be at some point—they aren't infinite. It would be possible only to extract coal in quantities that would last roughly for as long as the earth itself. There is no point in suggesting we do this, though, since we ought to have a little trust in the future. Yes, we can say, we are robbing the earth of all its coal, or rather robbing future generations. But they will invent something else to replace it. Something similar can be said in respect of the disadvantages of artificial bee breeding.

Yet it is still a good idea to be aware that what has developed in so wonderful a way in nature is disturbed by introducing a mechanical, artificial element into it. Since time immemorial, beekeeping has been seen as a wonderful pursuit. In ancient times people regarded the bees as sacred creatures. Why? Because, in their whole activity and industry they give us a picture of what happens within ourselves. If you take a piece of beeswax in your hands, you have there something somewhere between blood and muscles and bones. Inwardly in us these pass through a wax stage: this 'wax' does not become solid in the process but remains fluid until it can be led over into blood or muscles or bone cells. In wax, therefore, we have before us what we also possess as forces within us.

In olden times, when people made beeswax candles and lit them they saw this really as a sacred act. This wax that was being consumed had been harvested from the hive, where it was solid in form. As fire melts the wax and it evaporates, it acquires the same condition it has within our body. In the melting wax of the candle, people once intimated something within their body that flies up to heaven. This imbued them with a special sense of reverence, and in turn led them to regard bees as an especially sacred creature—since bees prepare something that we ourselves must continually prepare within us. The further back we go, the more we find reverence in people's attitude to bees and their whole nature. But at that time of course bees were still wild, and coming upon the wild colonies people of ancient times saw them as a revelation, only later domesticating them. Everything about bees is full of wonderful mysteries—you can only feel the nature of the bees if you carefully study what occurs between the human head and the body.

So these are my remarks for now. On Wednesday we'll have our next session, and there may be questions relating to this. Mr Müller himself may have something to add. I wanted to offer these remarks which are based on real knowledge, and therefore are beyond supposition. But it might be possible to make some things clearer.[6]

> It is worth studying the structure of this talk, which starts with artificial queen breeding, before broadening in scope. Artificial queen breeding is set against the subtle interplay at work within and around a bee colony. Steiner describes subtle formative and temporal influences on development, how the hexagonal form of the cell co-determines the life of the bee developing in it, and how the developmental periods of each type of bee have a relationship with cosmic influences, specifically the sun's orbit. Subsequently Steiner develops the themes with a clear inner dynamic, offering an overview, one might say, of the later lectures in the series.
>
> From the bees' activity in gathering nectar from flowers we move to its digestion and conversion into honey, the creation of wax, and the hexagon as the comb's structural form, and thence to the different types of cell and the bees that develop within them. Their developmental periods are related to their functions in the hive, and then more detail is given about the queen, the process of swarming, the action of bee poison, and the nuptial flight. Then follow observa-

tions on the human being, in which parallels between the human head and the hive are accentuated, and bees are compared to cells in the human body, as well as the influence of bees and honey on fertility.

After this survey, Steiner returns to his initial theme.

In the whole life of a bee colony, a bee organism, we find something, therefore, into which nature implants a wonderful wisdom. Bees are governed by marvellous natural forces of enormous importance; and therefore we feel a certain reluctance at intervening in or meddling with these things.

It remains true today that wherever human beings meddle with these natural forces they do not make things better but worse.[7]

Critique of modern beekeeping

In the final part of the lecture, modern beekeeping is subjected to critique. Rudolf Steiner refers to the life of the bee colony as an organism, implanted with a wonderful wisdom, saying that in their work and interactions with it beekeepers remove certain 'hindrances' and thus 'seemingly' make much of their work easier.

Modern beekeeping is now inconceivable without moveable frames. Fixed frames are rarely used any more. But we should not overlook the fact that moveable frames have encouraged a mechanistic view of bees. Anyone who works with them knows the contrast between the hive as a whole and its separate parts, and finding a conscious balance between the two is a challenge that has only emerged since moveable frames were developed.

Steiner's critique of modern beekeeping is much more pronounced when it comes to artificial rearing of queens. Since this involves making organic forces mechanical, as it were, he is concerned that this could, in the long term, weaken the ties between a colony and a purchased queen, so that breeding for ever higher yields of honey might result in increased susceptibility to illness. Nevertheless, he warns against being too fanatical in opposing this, suggesting that we cannot now put the clock back.

Biodynamic or Demeter beekeeping is one branch of modern organic beekeeping, and takes its lead from Rudolf Steiner's suggestions. It is distinctive in trying to ensure that interaction with the bee colony is founded on the idea of the intact organism: 'In accordance with biodynamic agriculture, beekeeping methods are oriented to the natural needs of the colony. Apiculture is organized so that the *bien** can give organic expression to its natural tendencies. In Demeter apiculture colonies can build their combs naturally. Reproduction, growth, regeneration of a colony and apicultural developments are all based on natural swarming. The bees use their own honey for their winter supplies.'[8]

'Natural' comb building means combs built by the bees themselves (i.e. not on wax foundations). Bees 'sweat out' of their bodies the wax they need for this, and form it into the comb. It is striking to see the glistening white, newly formed comb developing under a mass of bees, and it is a joy, as a beekeeper, to watch the growth of these organic forms. In a holistic view of the colony, the comb is a support organ like our human skeleton. If no partitioning frames are inserted, the bees create comb with their own distinctive distribution

* Translator's note: This word, taken from the German for bee (*Biene*) is now used also by English biodynamic beekeepers to refer to the super-organism of a whole bee colony.

of worker and drone cells. On each comb, the lower part, whether larger or smaller in size, is given over to drone cells. When the colony multiplies naturally through swarming, the beekeeper will try to make use of this instinct, the colony's own way of multiplying in spring out of a sense of abundance. Since it often proves impossible to catch a swarm, it is thought permissible to divide the colony, with its newly developing queens, into various units, thus forming new colonies. Biodynamic apiculture does not use artificially bred queens.

In the following lectures on bees, we find further practical suggestions which have been taken up by Demeter beekeepers. For supplementary or emergency winter feeding, they take care to add honey, camomile tea and salt to the sugar water.

3. SCIENCE

A noticeable aspect of the lectures to the workers is that Steiner often engages with the mainstream science of his day. At times he expresses criticism of it, which sometimes comes close to mockery, but at other places he displays detailed knowledge and admiration of scientific achievements. In studying bees he repeatedly urges us to employ precise, scientific observation. This frequently emerges, also, in the lecture cycle on bees. The following lecture focuses on questions relating to animal and human perceptions; and at the very outset Steiner refers to an article in the *Swiss Beekeeper's Journal* which asks whether bees see colours that are invisible to us.[9]

Now let's spend a few moments on this. These experiments done by Forel, Kühn and Pohl[10] show the thoughtless way in which such experiments are interpreted nowadays. You can scarcely think of anything more absurd than the interpretation given here! It is as if I had a substance (there are such substances) that was particularly sensitive to ultraviolet light, that is to the colours in the spectrum beyond blue and violet—for instance barium platinum cyanide, which is mentioned in the article. This shines when I screen out all other colours. So I screen out red, orange, yellow, green, blue, indigo, violet. In other words, first I have these latter colours, then I screen them—that is I cover them all up in the spectrum. And now I have these so-called ultraviolet rays which are invisible to human eyes. If I add this substance, a barium platinum cyanide screen, in the form of a white powder, and close the blinds, it begins to shine. We human beings see nothing in the darkened room; but now we allow these rays in, screen where they enter, thus only allowing in ultraviolet rays. But what happens when I add this barium platinum cyanide? Then, supposedly, it 'sees'.

You're doing exactly the same thing if you experiment on ants. Instead of using the barium platinum cyanide, you are now using ants. The ants head for sugar—and I therefore state that they see. In fact they do not need to see any more than the barium platinum cyanide needs to see in order to shine. All I can claim is that a substance exerts an effect on the ants. It is wrong to claim more than that. Thus the learned people who claim it are thoughtless, and assert things that are nothing but fantasy.

The only thing we can assert is this—proven, according to the article by the lack of any further results after painting over their eyes—that some kind of effect was exerted on these insects via their sense organs. But here, characteristically, the investigator simply transfers to ants and wasps what he has observed in bees—a thoughtless way of undertaking experiments.

Now we can also add this. If we pursue this further we come to what are called ultraviolet rays. First we have red, orange, yellow, green, blue, then indigo and violet too; here we have the infrared rays, and here the ultraviolet. So here, on the right, we have the ultraviolet rays, which, as the author himself puts it in the article, are distinguished by their very strong chemical action. In other words, what enters the field of ultraviolet rays is subject to strong chemical effect. And this means that if I introduce an ant here, it is immediately subject to this chemical action. It feels this. It's true, it feels it primarily in its eyes. This feeling is as if someone were fumbling at you when you come into the ultraviolet rays, in the same way that the barium platinum cyanide is affected when introduced to a chemical influence. If I screen out everything in a room and only allow in the ultraviolet rays, the ant immediately notices something's happening. Ant eggs, especially, will be changed completely under this influence—they would be destroyed by it, and so the ants rescue their eggs. In other words, this article describes a chemical influence. What I said recently is indeed true: I said that bees have a kind of smell-taste, something between the two. Bees—and ants are similar in this respect—feel such influences.

These gentlemen have so little idea of what is happening here that they do not know, for instance, that when a human being perceives colours through his eyes, and already when he perceives violet rays, small chemical changes occur. Colour perception in us is already oriented to chemical action. And so the whole thing investigated here in relation to bees involves the inner chemical change that occurs when bees live in ultraviolet light.

Now bees freely perceive everything in the realm of black, white, yellow, grey—which is only a darker white—and blue-grey. You see, there is no ultraviolet in these colours. And so these chemical effects that the bees feel so strongly when they come into a field of ultraviolet rays are not present in the above colours. When a bee leaves the realm of black, white, yellow, grey and blue-grey it feels something alien to it in the ultraviolet, a realm where it can do nothing. All depends on the fact that the bee has a kind of smell-taste. We usually distinguish very strongly between these two senses, don't we? Tasting is primarily a chemical sense, entirely dependent on chemical action; but the bee possesses a sense halfway between smelling and tasting.

This is not contradicted by the fact that a bee can distinguish a surface of paint on the front of its hive-box. You see, every colour has a different chemical action, and a different temperature. If you paint a red surface, for instance, and the bee approaches it, it feels this temperature. Naturally she will know the difference, therefore, between this and a blue surface. The blue surface feels colder. Thus the bee senses this warmth of the red, and the cold of blue. Naturally she can distinguish between them, but this does not mean we should assume that bees see with their eyes in the same way we do—it's utter nonsense of course.

The same is true of many other things people do. On a previous occasion I told you what all these experiments amount to. I spoke of a plant, the Venus flytrap which immediately contracts its leaves if you touch it. Just as you will instinctively make a fist if someone is about to hit you, so the Venus flytrap waits until an insect approaches, then closes tight. So people think this plant has a soul like we do. It perceives an insect approaching, and shuts its trap. Well, I always say that I know of something else that shuts suddenly if a creature approaches, and traps it—a mousetrap. So if we ascribe a soul to the Venus flytrap, we will have to allow the mousetrap a soul too. If we assume bees see ultraviolet light because they behave in a certain way when they enter its field, we will have to accord vision to barium platinum cyanide too.

If people would only think, they would discover some remarkable things, for barium platinum cyanide is highly interesting. Among other things it contains barium, a white metal that belongs to the potassium metals. Now, interestingly, such metals play a certain role in human life. If we didn't have such metals in our pancreas we could not properly benefit from the protein in the food we eat. We need these. In barium, therefore, we find something connected with the health of our digestion.

Platinum is a particularly valuable metal, as you know, also a very hard and heavy metal, a precious metal. All these metals are in turn connected with our capacity to sense and feel.

And now let us recall something else. This compound contains cyanide, a particular form of cyanuric acid, hydrocyanic acid. Now I told you that wherever our muscles are at work we always form some hydrocyanic acid. So this whole substance is similar to what we continually form in our body; and from this you can tell that human beings are particularly sensitive in their body—not their eyes—to what occurs in ultraviolet light, that is, in the chemical constituents of light. And so, by attending to such things, we ourselves can reach conclusions about them.

But only spiritual science enables us to attend to things such as these—that wherever barium platinum cyanide is especially volatile, a kind of feeling exists. And this is true of the bee to the very greatest degree. Bees feel colours with very special intensity, and only *see* colours shining weakly when a luminous living creature appears. That is why I said that in general the bees live in twilight. But when this new queen bee appears, she shimmers for them in such a way as glow-worms do for us in June. The three tiny eyes on the bee's head perceive this; its other eyes, the larger ones, do have a kind of light perception but as in twilight. And when this is dulled or screened the creature senses the presence of a colour with chemical action, ultraviolet, or one that has no chemical action at all, infrared.

At the end of the article in the 'Bee Journal' we read that more will be published later about infrared. Certainly, when bees enter an ultra-red field they behave very differently, for this no longer exerts a chemical effect. The actual experiments themselves are correct, but the conclusions drawn by Forel and Kühn are mistaken—they haven't applied thought to the results. People say they have proven something beyond doubt. Well, naturally they think so if they are willing to ascribe a soul to a mousetrap. But for anyone else who knows how far he can go or not, how far he can think in order to trace real causes, such findings are certainly not beyond doubt.

People are not generally accustomed to pursuing things with exactitude. Little things easily assume mistaken proportions in human appraisal, amongst academics and scholars also. Instead of staying with actual observations they let their thinking wander further and create fantastical things—elephants from fleas. When modern scientists assert something, they are wielding their power, and can only do so because all the journals are in their hands, and usually no one contradicts them. But in the end nothing will come of it all.

It seems to me that if you survey all of apiculture, you'll find that the best beekeepers don't bother much with the discoveries of Forel and Kühn. This is because they have to be practical, and will often instinctively do what is needed.

Of course it is better to know why you're doing something instinctively. It also seems to me that beekeepers may sit down on a Sunday evening, if it's snowing, say, and read an article like that because of course it interests them. But they won't be able to make much use of it, since it has no practical application. But now I'm sure you'll have some interesting questions.

Mr Müller: I'd like to say something about the queen bee. We discussed the fact that she lays eggs. But there are also unfertilized queens, for instance in bad weather, and from their eggs hatch only drones, of no value. But similarly, when the queen has departed and there is no young brood left, the worker bees raise a bee to be queen, and she too will lay eggs, but unfertilized ones, from which low-quality drones hatch out.

Then I'd like to say something about swarming. The first swarm as yet has no new queen, the new queen is pupating in her cell still. Only the older bees depart with the [old] queen. Then I take the queen out, and can bring the whole colony back into the hive.

As regards bees' vision, I'd like to say this: when we're working in the hive, and there is too much light—although still too little light for the beekeeper—the bees get terribly agitated. As regards bee stings while swarming: it is generally known that the bees are a little ticklish in a first swarm whereas they are less so in secondary swarms. In our view, young bees don't sting yet, they don't use their stinger. In some regions people don't harvest honey until the priest has blessed it—8 August is a honey day. It can also happen that the swarm flies off, the queen settles somewhere and it seems as if the swarm had finished; but this is not so, not entirely so.

Yes, as I said, it comes down to this: that the old queen departs when the young queen reveals herself, that is, when she shines out to the colony like a glow-worm. Once the swarm has left and you have caught the old queen, the colony can, as you say, be brought back into the hive, and takes up its work again calmly. This does not mean that the swarm did not leave because of the vivid light impression which the young queen emanates, perceived by the three tiny eyes. There is no contradiction here—you must proceed very logically. Let me cite an example drawn from life. Imagine that you were all employed somewhere and found you had to go on strike because of something done wrongly by the management. Let's assume you decide to strike, and down tools. You swarm off.

And now some time goes by and you find you can't buy any food. You're going hungry and are forced to go back to work. This does not mean, does it, that nothing wrong happened in the first place? You see, if you remove the old queen from the departed swarm, and bring the colony back into the hive, it will naturally find it has to put up with the new queen, bite the bullet if you like, because, as it senses, it no longer has the old queen. So what I said was not wrong; one just has to look at everything in the right light.

Then you spoke about first swarming, before a young queen has appeared, and so, you thought, she can't have an influence. Now have you ever observed a first swarm before an egg has even been laid in a queen cell?

Mr Müller: Nine days before the young queen hatched.

Initially the young queen is in her cell as an egg, and is fully grown as a queen after 16 days, when she hatches. Nine days previously she would have been developing already. Now the remarkable thing here is that the egg glows most strongly of all. The young queen still glows for a while, then ceases to. But she glows most strongly when still an egg and larva. So it's quite understandable that first swarms happen then, composed of the most sensitive bees. Nothing happens before a young queen is present, even if only as an egg.

If the queen remains unfertilized, she doesn't produce workers but only drones and, as Mr Müller told us, poor quality drones at that. This is correct. Such a brood, from an unfertilized queen, known as a virgin queen, is useless since it contains no worker bees. It is therefore important to ensure that the nuptial flight can take place under the sway of sunlight.

Again you can see what an important role is played by chemical factors. Everything occurring here acts upon the bee's sexual nature, which is entirely chemical in nature. When the queen bee flies high in the sky, naturally the influence upon her is not so much from light as from the chemical elements in this light effect. Here, specifically, you can see how subtly sensitive the bee is to chemical influences.

You also said that beekeepers working with the hive need light, which makes the bees restless. Now picture this as vividly as possible. Through light the bee receives chemical effects which she senses very strongly. When you come along and let light into the hive, this is like a sudden strong gust of wind for the bee—like opening a window and letting in a sudden strong draught. The bee feels the light; she does not feel that it is particularly bright, but feels this as a shock, is greatly unsettled. And, though I haven't seen it myself, one could say that when the beekeeper does this, letting in a lot of light, the bees behave very nervously, become inwardly restless, succumb to these chemical effects of the light, and begin to fly up and down like little swallows. They dance back and forth, a sign that they feel inwardly agitated. Bees would not behave so nervously from seeing only light itself, but would just tend to crawl away into a dark corner to escape it.

Now in all these matters we should remember of course that there are effects involved here that cannot be compared with how things affect us as human beings. There's a danger that we anthropomorphize everything, believing that because we see in a certain way creatures or insects also do. That shouldn't be assumed. You may perhaps have observed the following. Imagine you're in a kitchen and the stove has been lit. The cat, who likes sitting on top of it when it's warm, curls up and sleeps with her eyes shut. Now if there is a mouse somewhere under a cupboard, though she cannot see it, she may suddenly leap down and catch it even without opening her eyes; her instinct is unerring and she has the mouse in her jaws before you can say Jack Robinson.

Now, the cat didn't see the mouse, for her eyes were shut, she was asleep. Well, people say that she has excellent hearing, and heard the mouse make a sound. In that case you'd have to claim that the cat hears best when asleep, which is a somewhat dubious proposition since sight and hearing only come into their own in the waking state, whereas smell, for instance, does play a very important role during sleep. It acts chemically, and something chemical occurs in the nose and the whole brain. But even if you were to hear something when asleep, could you leap on it with such unerring accuracy? Certainly not. Hearing is not so exactly targeted. So we can rule out hearing here. But what we do find in the cat is a very fine and subtle sense of smell, which she has in her bristly muzzle. And this very fine sense of smell is due to the fact that in every such bristle is a channel containing a chemical substance which alters in the presence of the mouse. In the absence of a mouse, the substance has a certain chemical property. If there is a mouse in the vicinity, even a good way off, the cat discerns this through the chemical response in her bristles. I once told you about people who live, say, on a third floor but can detect a substance, for instance buckwheat, that is being stored in the cellar. It can make them feel ill. It is clear that the sense of smell is very reliable, otherwise you couldn't have police sniffer dogs. They accomplish a great deal through their sense of smell, rather than their eyes. In the animal kingdom we should ascribe acuity of the senses not to the eyes but to chemical effects, most strongly so in the case of ultraviolet rays.

If you were to accompany a sniffer dog, taking with you a screening lantern so that he was always bathed in ultraviolet light, the dog would find what he is looking for with still more certainty: his olfactory hairs would detect the chemical effects still more clearly.

Thus everything we can know about animals involves discounting the more conscious senses, and delving further down into the sense of taste and smell, that is, into the chemical senses.

And then you thought that young bees do not sting. But it is very understandable that young bees do not yet have a stinger, have not yet developed their whole inner organization to a sufficient degree. This only develops as they grow older. There is nothing special in this and it does not contradict what I said.

Mr Müller asks about artificial feeding, saying that he takes four litres of water, five kilos of sugar, adds thyme, camomile tea and a pinch of salt. What kind of effect might this have?

This area is one we can offer insight in quite particularly since our own medicines are based on principles such as these, though here undertaken instinctively. Not all our medicines but a number of them are based on such principles.

Feeding sugar to bees is, as such, nonsense really, since by nature bees do not feed on sugar but on honey and pollen.

Mr Müller: Sometimes for instance you have to remove the forest honey stores,[11] *and even the brood combs, because otherwise the bees get dysentery. And then they may be left with only two to three kilos and that isn't enough for them.*

In general bees are not accustomed to eating sugar. Their proper food is honey really—that's what they're naturally used to. Now it is remarkable that in winter the bee converts every food it is given into a kind of honey. In other words, the bee transforms the food it assimilates. As it digests its food, it is capable of turning it into a kind of honey that it consumes during the winter. You can understand that this requires more energy of bees than when they consume honey. Then they don't expend the energy they must otherwise find to convert sugar into honey in their organism.

What kind of bees will extensively convert sugar into honey themselves? Only the strong and therefore useful ones. Weak bees cannot be induced to convert sugar into honey, and therefore they are more or less useless. Now earlier I said that in our approach we can easily explain why you would add camomile tea to the winter feed, for instance, for by doing this you relieve the bee of some of the work she must otherwise do inside her own body. If you add camomile tea to the sugar, this camomile substance is from the part of the plant that produces honey. You see, the substance contained in camomile tea is not just contained in camomile, but in every honey-producing plant. Camomile, however, contains this substance in greater amounts than other plants. A plant contains a great deal of starch, which has the tendency to turn into sugar. In the camomile plant, the sap acts on this starch in such a way that it already begins to convert the syrupy sap into a condition close to honey. Thus if you give bees camomile tea, you support the honey process in them. You already make the sugar into something approaching honey if you add camomile tea to it.

We do the same with our medicines. You cannot simply give a metal to someone as a remedy since they won't be able to assimilate it. You have to add something that makes it easier to assimilate. The same is true of the sugar to which camomile tea is added.

Salt has to be added too, since in general it makes things that are otherwise hard to digest more digestible. We instinctively add salt to our soup, since it has the property of swiftly dispersing through the body and helping us digest our food.[12]

> This lecture starts from bees' sense perception. Rudolf Steiner does not dispute that UV rays can call forth a sensory response in these creatures, he is only unhappy with how such experiments are interpreted. He does not accept that our own experience of sight should be assigned to bees, since this projects a quality of human experience onto them. While bees respond to optical phenomena, according to Steiner the sense perceptions triggered are different from our own. Other animals—Steiner refers to cats and dogs by way of comparison—live strongly in a world of smell and taste. Bees spend most of their life in the twilight obscurity of the hive, and are strongly dependent on orientation governed by smell, taste and touch.
>
> We know today that bees have an outstanding sense of smell, as tested in learning experiments. According to Jürgen Tautz,[13] they have an extraordinary capacity to learn about scents: brief contact with a fragrance is enough for bees to recognize it again with 90 per cent certainty. After two to three positive contacts, they can recognize the fragrance again without error, while colours and shapes require longer training. In their search for coloured blossoms, fragrance plays a primary role too. It has been observed that, in their quest for a nectar source, bees fly against the wind and thus towards the scent streaming from flowers.
>
> The investigations by Professor Karl von Frisch[14]

demonstrate that the way bees see colours does have a certain resemblance to our own. The main difference lies in their insensitivity to red, and their extraordinary sensitivity to ultraviolet. We really cannot imagine what they actually perceive in their colour world—whether they *see* or *feel* colours. But we know what it means to *feel* the effect of colours; and this comes to expression in our symbolic language. Red, a colour which actively leaps out at the viewer, is thought to be 'warm'—as we see on the hot tap or in warning signs, or the red traffic light. Blue stands for coldness, and invokes a sense of breadth, distance and yearning. In relation to bees, Steiner speaks of a 'cold-warm' colour perception: towards ultraviolet a chemical response is elicited, and towards infrared warmth is perceived.

The 'glow' of the growing queen is described as another perception by Steiner. It is hard to know how to understand this. He describes the three tiny eyes on the head of the bee as sense organs for this perception. Nowadays it is assumed that these tiny ocelli perceive light and dark conditions. They should not be confused with the bees' large compound eyes. However, the 'glow effect' can scarcely be thought of in optical terms. We can be sure, however, that the development of new, future queens is an exceptional condition in the hive since it initiates the loss of the hive's hitherto intact unity. As soon as the first new queens develop, the old queen, at the centre of that unity, is no longer alone.

And we can also easily understand that the colony's unity and identity could be connected with bee poison, although as yet there are no trials on this. The poison is a defensive substance which the colony uses to repel invaders, and whose smell triggers alarm. Rudolf Steiner's descriptions lead us to suspect that the bee poison inwardly consolidates unity. Reduction of bee poison within the hive would then let this unity lapse—a necessary precondition for the

> imminent division of the colony. It must prepare through physiological changes for these unusual circumstances that arise in the swarming period. The bees gorge themselves on supplies to strengthen themselves for this challenge. The glowing of egg and larva of a new queen could be understood as a kind of 'clairvoyant perception' which announces the imminent swarming process.
>
> The accusation that scientists are guilty of anthropomorphism is a telling one. Rudolf Steiner considers it wrong to foist our human way of seeing on another creature. He believes that one of our central tasks as human beings is to use our thinking in scientific discernment, and he repeatedly stresses the importance of studying how our organism is related to the natural world. The human capacity for knowledge and experience can lead us to discover how natural agencies are also at work within us. Thus by observing ourselves carefully we can sense how natural forces work, manifesting spiritual principles on which the world is founded. Only the human being is capable of consciously comprehending this spirit that underlies the world; and such perceiving discernment gives rise to a different connection to the world. It requires us to stand fully in the world and not merely observe it neutrally, from a distance, or merely measure, calculate and quantify it.

We saw that the inmost core of the world comes to expression in our knowledge. The lawful harmony that governs the world manifests in human knowledge.

It is thus part of our calling as human beings to *manifest* in reality the fundamental laws of the world which govern all existence but would never otherwise come into existence themselves. The nature of knowledge is to encompass the ground of the world never found in objective reality. In metaphorical terms we can say that our cognition is a continual entering into the ground of the world.[15]

Foundations of science

Before embarking on the rest of the lectures it is helpful to consider in more detail what Rudolf Steiner means by 'scientific': where he sees the limits of modern science, but also what relationship we develop to science today as transmitted by our schooling.

At the beginning of his career, as his early works testify, Rudolf Steiner was intensely preoccupied with the writings of Johann Wolfgang Goethe. He was struck by the way Goethe approached nature and lent his findings artistic expression. He saw in Goethe's scientific and philosophical works, especially, a means of redressing the deficiencies of the prevailing scientific outlook at the time. In brief summary we can identify three fundamental impulses at work in Goethe's world-view:

1. All things manifest as a unity of spirit and matter. Only the human being himself separates phenomena into outer and inner aspects. Through observations we experience the world and can thus perceive the outward aspect of things. The inner aspect—that of ideas—has to be added through our own activity.
2. Goethe seeks a methodological diversity; for in trying to grasp life processes, the procedure we use cannot be the same as when we study physical laws.
3. Nature, as an artist, continually elaborates and develops her natural processes, bringing forth phenomena from inner potential. In a similar, creative process, the human being can grasp her laws through scientific discernment, and embody them in artistic form.

Thinking has the same meaning in relation to ideas as the eye does to light and the ear to tone. *It is an organ of perception.*

This view is able to reconcile two things that today are usually thought completely irreconcilable: empirical method and scientific Idealism. It is thought that accepting the former means dismissing the latter, but this is not right at all. Of course this is what people inevitably think if they regard our senses as the only organs for perceiving objective reality. The senses only offer us contexts that can be traced back to mechanical laws, and thus the mechanical world-view would be the only true one if these assumptions were true. But here people make the mistake of simply overlooking the other equally objective constituents of reality that *cannot* be attributed to mechanical laws. Objective reality is not confined to sense reality as the mechanical world-view believes. Sense phenomena are only half of the given world. The other half of it are ideas, whose organ is thinking. Albeit of a higher order, they are also an experiential reality.[16]

> More so even than in Rudolf Steiner's day we now live in a world informed by the scientific world-view, founded on mechanical laws and on what can be mathematically calculated. In modern biology, we have seen an unimaginable refinement of knowledge about chemical and molecular laws. Those who engage with this can be astonished at the complexity and diversity of all that is described. Research has been conducted into the biochemical processes underpinning life, and life processes can be understood in terms of chemistry and physics. And yet life itself has still not been explained. The nature and quality of living things remains an enigma, and questions about the whole context of life have not been fully answered. No research has been done into life's irreducible uniqueness, and for many it is clear

that the modern scientific approach is one-sided, revealing only half of reality and leaving the other unanswered.

Darwin's theory of evolution is a good example of this. The way in which it explains the evolution of life is persuasive, offering a complex, carefully elaborated structure of thought that conceives of a directionless process governed by random occurrence. But doubts can arise here: while much of Darwin's view can be true, it remains unsatisfactory as a sole explanation of how and why the life forms we know today evolved. Surely there is a force that governs all evolution. Can everything really be founded on past causation? Surely there can be causative factors which we might, in the broadest sense, call the goal or aim of evolutionary processes. Modern ideas about evolution need to be enlarged by a broader, complementary outlook. Such questions relating to Darwin's theory were no less topical in Steiner's day.

People think they can attribute the character of a particular species to the external conditions in which it has been living, in the same way, say, that the warming of a body can be explained by the sun rays falling on it. But they completely overlook the fact that the inner laws of this character can never be the consequence of such conditions. Environmental factors may have a determining influence but they are not an *original* cause.[17]

Rudolf Steiner consistently sought to include what he calls the other half of reality. This is not somewhere far distant but immediately present in all sense phenomena as well as in chemical processes. Thus when Steiner describes the spiritual dimension of many substances, he presents only the other side of material reality.

Ordinary experience is only half of reality. The senses only relate to this half. The other half is present only for our spiritual capacity of perception. The spirit raises experience from a 'sense phenomenon' to its own domain. We have shown how, in this domain, it is possible to raise oneself from the created world to a creating one. The *spirit* finds the latter when it approaches the former.[18]

Science in the modern age is a sensory science, and has therefore taken up only a circumscribed part of the world into its evolutionary doctrine. Just as your body is not the whole of you, so neither is matter the whole of the world. Just as life, feelings, thoughts and drives exist in your body, and are invisible to bodily eyes, so the spirit is likewise a reality in the world.[19]

> In his lectures on bees, Steiner speaks a great deal about silicon dioxide in its mineral form as quartz or silica, describing processes in the world, in minerals, plants, animals and human beings. In many examples, he describes the spiritual qualities of the physical and chemical substance silica, speaking of forces that become active through this substance, and also of how silica mediates or governs a relationship to the spirit. In these life processes and movements, the dynamic aspect of silica becomes apparent. Quartz is thus only a dead, end-stage of a living silicic acid process.
>
> Often we meet these accounts with astonishment, lacking access to the faculty of experience that would enable us to trace these processes ourselves at first hand. Reproducibility is the basic stance of every scientific outlook. But there is a way forward. We can bring our own feelings to bear on these accounts drawn from spiritual vision, and experience their

truth in the same way as we experience the truth of an art-work. Something in us responds, intimating as yet hidden reality. Even better, we may sometimes rediscover insights already germinal in us in an account that we read by Steiner. But even when we gain a sense of truth from Steiner's spiritual descriptions, we may have little or no trust in such felt reality since we mostly have not developed our feelings into an objective organ of perception. By contrast, it seems that the intellect, inculcated in us from infancy, offers us certainty and objectivity. And then we may only be able to follow Steiner's accounts with our abstract thinking—which sends us back in turn to the world of mechanical thought, excluding us from deeper comprehension.

This digression on the scientific world-view will hopefully help us understand Steiner's lectures on bees. Steiner is, for instance, not concerned only with the biochemical properties of quartz or formic acid, but with two important natural forces that play a part in a healthy or ailing environment. In his accounts of these processes he describes, above all, the other side of reality, the inner lawfulness, the aspect that only human beings can comprehend. And thus we can understand his appeal to us to acquire a feeling for the life of bees by seeking to perceive what also acts within us.

Everything about bees is full of wonderful mysteries—you can only feel the nature of the bees if you carefully study what occurs between the human head and the body.[20]

4. THE EFFICACY OF HONEY

Mr Müller has given me another issue of the *Swiss Beekeeper's Journal* with an article about honey cures: 'Our Experiences with Honey Cures at Frauenfelder Children's Home' by Dr Paula Emrich.[21] [Passages from the article are read out.]

Some remarks spring to mind in relation to this interesting article, which describes efforts to treat children in this home with honey. The children were all weakened in various ways by poor nutrition. As described here, they were given honey dissolved and finely dispersed in warm milk—milk which is not overheated, not boiled. It was kept below boiling point.

They had outstanding results. In particular the author of this article found, very happily, that the red blood corpuscle count in the children increased to a very great degree. For instance, they had two children who were siblings.[22] The younger had a red blood corpuscle count of only 53 per cent when she was admitted to the home. After discharge, that is, after she had taken the honey cure, this figure had risen to 82 per cent. The older child's red blood corpuscle count was 70 per cent, and this rose by the end to 78 per cent. So in the latter case the increase was less, but there was still an improvement. The older child was only given milk, not honey; this child wasn't so weak to begin with, but the improvement was not so pronounced either.

She cites a whole number of very interesting trials here. It is important to note the age of the children in each case. If you want to study the effects of a substance on human beings, it is no good at all simply running tests in a lab. You must always also first establish the patient's age—something invariably done first at a consultation with any patient—when undertaking nutritional or medical remedies.

So here we have an eleven-year-old boy who had honey treatment for eight weeks. This greatly improved his glandular condition, and also apical catarrh. The red blood corpuscle count, which is significant, rose from 53 to 75 per cent. Then we have another eleven-year-old boy, whose count rose from 55 to 74 per cent. Then a 14-year-old girl whose count increased from 70 to 88 per cent. I won't go on—the other cases showed similar improvements. The author cites weight increase too, which also demonstrates that the children grew stronger. Then she cites the following cases: a ten-year-old girl, another ten-year-old girl, a 13-year-old boy, a seven-year-old girl, an eleven-year-old boy, an eight-year-old boy, a twelve-year-old boy, a nine-year-old boy and a seven-year-old boy. The trials show that children of this age—let us say broadly school-age children—greatly benefit from a honey cure.

The author also seeks to discover why honey is so very beneficial for these children. And here she says something very interesting, which is at the same time a serious condemnation of the kind of science so widely resorted to nowadays.

What do scientists do today when they want to find out the nutritional value of a food? They analyse its various substances and investigate what chemical constituents compose it.

Now the author says that a pupil of the famous professor of physiology Bunge,[23] whose name I'm sure you've heard, was in Basel and undertook experiments, feeding milk to mice. These mice did very well out of it: they developed very well.

But then he tried the experiment in a different way. Milk, he thought, consists of casein, which can be broken down in turn into fat, sugar and salts. And so, he thought, since the mice were thriving on milk, I will give one group of them the constituents of milk—the fat, sugar and salts which make up casein. That is the same as milk contains. But the mice who were given this died after a couple of days! They got the same substances, but they died. So you can see that material substances alone do not account for everything. Something else must be involved. At least, that is the conclusion these scientists should have come to.

But instead, since they only acknowledge material substances, they assumed milk must contain another substance too, and that any benefit of milk must be down to its substances. If casein, composed of fat, sugar and salt, is not showing this benefit, they thought, there must be a further substance in milk, in such tiny quantities that chemical analysis does not reveal it. And now these folk named this substance 'vitamin'.[24] 'Vita' means 'life', and vitamin means 'life-maker'.

The poet Heine once mocked this kind of thing: there are people, he says, who try to explain where poverty comes from, and do so by saying it comes from penury.[25] In other words, they simply find another word for the same thing. In fact it explains nothing.

I once attended a discussion on the origins of comedy, at which a number of people expressed some fine thoughts on this theme—where comedy comes from, what makes us laugh. Then someone stood up and took the floor—and you could see by his demeanour that he had something very important to say! He presented his view of comedy by saying that comedy derives from the fact that human beings have the *vis comica*. 'Vis' means 'force', 'comica' means 'comic'. So we have the 'comic force' and that explains it! That is like saying money comes from the money-making force—it tells us nothing.

Economists would not be satisfied if someone told them that money originates in the money-creating agency. But in science people do not so easily discern such things. Where does the enlivening power in milk originate? In 'vitamins'. Well this is like saying poverty comes from penury. But they don't notice the tautology of it. They think they have said something very important, but they have said nothing.

This is the irritating thing about modern science. People believe they are saying something important, and they announce it in grandiose words—and other people believe everything they say. But if this continues for any great length of time, things will come to a dire pass. The world depends on us being able to do something, not on mere talk and empty words. Words must signify something real. In the past there existed a science directly linked to practice, while today science with no practical application at all is often pursued. It simply plays with words, and exercises a new authority that has joined the old ones.

Not so long ago there were fewer specialist journals than today. Information and articles we can now read in these journals used to be shared between people at conferences—beekeepers at apiculture gatherings for instance. I can remember from my youth how people consulted each other at beekeeper gatherings: people would share their experiences, and you could tell straight away if someone was just an old gasbag or actually knew what he was talking about. It's quite a different thing if you actually hear someone speaking—you can tell if he knows anything or if he is simply regurgitating what he has heard elsewhere. By contrast, print in a journal gains a kind of authority that may or may not be deserved. It has in many respects joined the other forms of authority. If something is printed people believe there must be some truth in it.

But in this case, in this article, there is something to it. This physician has achieved great benefits through her honey cures. Her practical work is really excellent. But reflecting on her results in terms of science, she does not get anywhere really. She herself acknowledges this:

> It would be desirable for our findings to be made known in the broadest circles, and especially for more honey to be given to growing children[26] [...] For now, we simply present the results of our practical trials; as the science of vitamins develops, no doubt pharmacologists and physiologists will turn their attention to the action of honey on the organism.[27]

Likewise she says at the beginning of her article:

> I am very keen to report on the effects of honey cures from a medical point of view [...] Our success encourages us to try to discover the underlying reasons for these benefits. Although I am aware that I have scarcely begun to get to the bottom of the matter, I would like to offer my experiences and test results as a point of departure for further research.[28]

Thus it is clear from her own words that she is humble enough so recognize that the whole vitamin theory cannot really fathom what is at work here.

But now let us consider the following and explore what underlies this honey effect. The trials cited show us something, you will agree. They tell us that the action of honey is especially strong—and further trials will bring this to light still more clearly—not in infants but in children who have either reached second dentition or are a good deal older than that. The trials themselves tell us this. It is extremely important to take account of this. But the trials show something else too: that honey produces the best results if added to moderately warm milk. A honey-milk mix is especially beneficial for children.

And if we were to explore this further we would discover the following: that honey can be important also for younger children, but then only a little honey should be added to the milk—more milk, less honey. In the elderly, honey is particularly beneficial, not milk. Good results are obtained in elderly patients by giving them honey on its own, without milk.

Milk and honey, it has to be said, are very important for human life; and these findings show this.

And the old sciences, as I have often said, were not as useless as modern scholars think. They may sometimes be couched in simple terms but they are actually very insightful and wise. You know the old phrase: a 'land flowing with milk and honey'—which means a healthy land, a place where one can live in health. Well, people in the old days knew that milk and honey are strongly connected with life.

Nature sometimes speaks to us very clearly, and we can understand what she says if we attend properly to simple matters. If you know that nature in general is very wise, you will not need much proof of milk's benefits for young and growing children. Otherwise honey would flow from mothers' breasts and not milk. This is by no means an impossible thing to imagine since plants produce honey and it would be quite possible for women's breasts to produce it too. You need only think in simple enough terms. Instead of thinking that nature is incompetent because it only creates mother's milk, not mother's honey, you can realize the wisdom in this: that an infant needs milk above all, and that honey can be added as a child grows older.

It is not adequate to deal in mere word play—to say that poverty comes from penury, comedy from the *vis comica* and the enlivening power of honey from the vitamin it contains. Instead, let us study reality; and when I say the following to you, it is something you are by now fully aware of from these lectures, but we should always look at things in the right light.

When you travel to a high, mountainous region, you will find quartz crystals wherever the mountains are hardest, where, in a sense, the hardest earth element is exposed. They are very beautiful. You can find all kinds of crystals there. You'll remember that I drew you a picture of these quartz crystals—they look like this. If they are whole they are closed off at both bottom and top, but usually they aren't whole and intact. They emerge from the rock, in a sense grow forth from it in the shape I have drawn here. What does this mean? The earth grows hexagonal crystals like this, ending in an apex. So the earth possesses the force to shape something hexagonally.

And, as I have often said, we contain within us the forces that are in the earth and the cosmos too. The earth acquires this force from the cosmos, and we acquire it in turn from the earth. We possess within us the same force that drives quartz crystal forth from the ground. You see, the human body is actually full of quartz.

And now you'll say, 'For heaven's sake, what is he talking about?' Quartz, as we find it in the high mountains, is one of the hardest substances. If you bashed your head against it, your head would break before the quartz did. So the most striking property of quartz is its hardness. But substances do not take the same form in different conditions. We contain within us a substance identical to quartz but in a more fluid form. Why so?

You see, one can observe—through true inner vision—how something continually streams down from the human head into the limbs. This is very interesting. Here is the human head, and now streaming down continually from it we find the same substance that the earth once expressed from within, which became hard and was, for instance, deposited as quartz crystal. It streamed out from the earth's interior; and in us it streams down from the head into the whole body. This is quartz or silicic acid. But the human body doesn't allow the quartz to crystallize. It would certainly be a fine thing if our insides were filled with nothing but quartz crystal. That would be painful indeed. The human body lets quartz get as far as growing hexagonal, and then it stops it. It doesn't let it go any further. So in our body we have only the beginning of quartz formation, and then it stops. And our life depends on us continually trying to form hexagonal crystals from what streams down from the head, but then preventing this, making it cease. Inside here, such crystals continually try to form but are then prevented before they get as far as that. We contain this 'quartz sap' within us, in high dilution if you like.

If we didn't have this, however much sugar we ate we would never taste sweetness. The taste of sweetness is made possible by the quartz within us—not by its materiality but by virtue of its will to take hexagonal form, to become crystalline. That is the important thing here.

In the earth, therefore, the same process exists, but is led further. We prevent the silicic acid from going any further when it is about to start crystallizing in us. The earth lets it consolidate upwards, crystallize. But we need this silica force which is the same as produces hexagonal forms. The power to produce hexagonal forms is one the human being needs.

I'm sure not all of you are good at geometry, or you're not all equally well versed in it. You may not be able to take pen and paper and draw a quartz crystal yourself, or model it in plasticine. But your body is a good geometrician and continually wants to create such crystals. We are prevented from doing so; but in fact all life involves us impeding the process of dying, and when we no longer do so then we die.

And now let us consider bees. The bee flies out, collects nectar, then assimilates it within her and draws her own life forces from it. She also creates wax. And what does she do with the wax? She forms hexagonal cells. You see, the earth makes hexagonal silica crystals and the bee makes hexagonal cells.

This is terribly interesting. If I were to draw the cells of the bee for you, as Mr Müller showed them to us, they look like the quartz crystal, except for the fact that they are hollow. The quartz is not hollow. But in form they are identical.

Yes, these cells are hollow; and what is placed inside them? The bee egg. Whereas quartz contains silica, the bee cell is hollow, and into this is laid a bee egg. And so the bee develops by virtue of the same force active within the earth and forming quartz, a process in which finely distributed silica is at work. There is a power at work here that cannot be physically demonstrated. And honey works through the bee's body so that it can shape wax in precisely the form that the human being needs, for we have to have these hexagonal spaces in us. We need the same. And as the creature that can best configure these dynamic hexagonal forces, the bee gathers from everything present the food that can best be led over into these same dynamic forces in the body.

If you eat honey it provides you with a hugely strengthening force. If you have grown too weak to develop within you these hexagonal powers that have to flow down from the head into your whole body, and thus no longer have the strength to give the blood the stability that ensures these hexagon forces are continually available, then honey is required or, in the young child, milk. The young child does not yet have these hexagon forces, and must be given them therefore in the milk that is prepared within the mother.

And so you can give endless amounts of casein, fat, sugar and salts to mice, and they still die. Why? Because they also need these dynamic hexagon forces. If you just mix casein, fat, sugar and salts together, this mixture does not contain the forces active in hexagonal form. But if you give mice milk, this contains them. They are not strongly enough present in milk to form hexagonal crystals when the milk goes sour. If these hexagonal forces were a little stronger in milk, you could drink sour milk that formed little silica salts on the tongue. It would seem then as if there were lots of tiny hairs in the milk. But the process does not go far enough in milk since it comes from the human body itself, or the animal's body, and remains fluid there. This is sufficient for the child but no longer for the adult; and the path to adulthood already begins of course in childhood. And here stronger dynamic hexagonal forces are needed, such as honey contains.

This is very interesting: milk from a child's mother is actually animal-like in nature—has an affinity with animal nature. But honey comes from the plant world, via bees. It comes from the plant kingdom, is plantlike in nature. If you take silica, thus quartz, it is mineral, and has a clearly hexagonal form. The wax that develops in the bee herself as a consequence of what the bee eats, acquires the same form; it does not arise from this form but is given it in the hexagonal cell. Milk dissolves this form again in turn. In the milk there arises only a shadow image of hexagonal crystals. And so we can say that honey is something that we must welcome most heartily of all.

One might think, of course, that it would be a good idea for people to eat silica instead of honey and in this way assimilate the hexagon force. But by virtue of being driven into hexagonal shape, of acquiring six-sided form, silica exerts too strong an effect, though it is, nevertheless, something beneficial.

Let us imagine the following. Think of a child living in poverty who is not thriving, and is given a honey cure as described in the article, at the age of 16 or 17, or at 13 or 14 when it would be best. The child is ailing, and the percentage of iron corpuscles in his blood is continually decreasing, weakening. The child grows up, reaches, say, the age of 30, is still living in poverty and has grown very weak. The author of this article also describes these circumstances, saying that such people collapse. Someone who reaches the age of 30 under these conditions could of course perfectly well be given a honey cure, but he is now too emaciated. He would have to eat so much honey to gain any benefit that it would, in turn, wreck his stomach. Honey, you see, is at the same time something that requires moderation. Too much of it wrecks the digestion.

There is a simple reason for this: honey is sweet, contains a great deal of sugar, but the stomach chiefly needs acidity; and if you introduce too much sweetness into the stomach you spoil its acidity. Only modest amounts should be taken. If someone were so emaciated by the age of 30 and therefore required large amounts of honey for the cure to help him—which it certainly would do—this would on the other hand cause him stomach problems, and intestinal disorders. So it cannot be done.

But we can do something else instead. We can first give him highly dilute, powdered quartz, silicic acid in other words, as a medicine. And after a while this will enable him to enjoy the benefit of small quantities of honey. The highly dilute silica will have activated the hexagon force within him, and this can be followed by a smaller amount of honey. Silica paves the way for the honey.

An emaciated 30-year-old, whose haemoglobin count is low, can also be given the honey mixed *with* the highly dilute silica—rather than with milk as in young children. This helps the honey to take effect.

One has to know such things. One has to ask how honey acts on us, and recognize that it is through the hexagon-forming agency that exists in the bee, as we can tell from looking at her wax cells. And that is why honey is so beneficial. In the child the strength of milk acts primarily, but this can be reinforced through honey, while in the adult the action of honey is primary. Once a person is older, you have to reinforce this honey action with that of quartz, as I said. But a honey and milk cure can also help, since a person still has the forces of early childhood within him; and a simple honey cure can help too. There is no disputing the benefit of a honey cure.

This is well known, and used in practice. It is just a question of ensuring that people know about using the right amounts of honey; and of course they are easily misled, easily deceived by the prevailing culture. You may have found sometimes on your travels that when you asked for honey in a hotel what you were actually given was syrup instead, an artificial product made from sugar. If people realized that there is a real difference, that this crystallizing potency is simply lacking in syrup, they would know that it cannot have the same effect as honey. You could of course also feed mice on honey, and they would find it very tasty. But syrup, artificial honey, would soon finish them off, even though they might survive on it for more than a few days.

So that is what I wanted to say about this article on milk and honey cures. And now something else of interest has reached me. I'd like to speak about it, and hear what you think, and what Mr Müller will have to say about it. I think this will lead to a great many questions, and so we should carry on discussing it next time. Then you can put your questions, and either I or Mr Müller will answer them. For now I just want to briefly mention two things. You will consider them odd but I am very interested to see what you have to say in response.[29]

> The article read out concerns the action of honey-milk cures on children, and leads to questions about the effect of such a cure. Steiner cites vitamins as an example of a misunderstood action. His view is diametrically opposed to this outlook. Scientific research into vitamins had begun in 1910 and regarded vitamins as essential substances that the body cannot produce itself. The term is composed of the Latin *vita* (life) and *amines* (nitrogen compounds). By contrast Steiner is seeking for the dynamic life, that real 'vita', which in his view is sustained by the action of honey.
>
> We now know that chemical analysis of the constituents of bee products does not help us understand their health-

promoting properties. The latter arise from an interplay of diverse active substances. This has led to a new branch of modern medicine, apitherapy. There is a long tradition of research in this field, chiefly in the countries of eastern Europe and China; and it is now meeting with increasing interest in western Europe. Bees provide us with six 'products' which all have proven benefits. Besides honey, these are pollen, propolis, beeswax, royal jelly—the food on which queens are raised—and bee poison, whose medical use will also be discussed in the following lectures by Steiner.

This lecture focuses on the action of honey, which is first described in relation to trials with children. Then Steiner explains the silica process in the body. His references to the 'hexagon force' and its effect on us are not easy to understand, since this silica process is hard to grasp empirically, through observation. But the frequent recurrence of this theme shows crystallizing forces to be a key concern of the lectures. Complementary passages from other lectures may help here to lend clearer definition to this picture of silica forces and their action.

Bees, you see, gather from flowers what was originally present to make the hexagonal quartz crystal. Bees draw this forth from the flower and, through their own body, make replicas of the quartz crystal. Between the bee and the flower something occurs that resembles what once occurred out in the macrocosm.[30]

Nectar formation in the plant is closely connected with light and warmth. Sugars formed there arise through the process of photosynthesis from air, light and water. Thus nectar is a substance as yet scarcely terrestrial, and accordingly tends to evaporate quickly again. Bees are capable of harvesting this

> quality and making use of it in their life. The nectar from the flower is processed by the bees who enrich it and make it durable with substances from their own bodies. By virtue of this assimilation and conversion, bees can build their wax comb structures. Here a modern bee researcher, Jürgen Tautz,[31] tells us that the hexagonal cells are not formed by the bees through their experience or intelligence but that the capacity to create hexagonal structure is, rather, inherent in the wax itself. Through the engendered warmth necessary for building the comb, the wax directly acquires its hexagonal form. The bees therefore exude a substance that already bears the 'hexagon agency' in itself. These hexagonal cells are in turn the cradle for new generations of bees, as well as a storage facility for their honey. At various levels, therefore, we can discern the connection between bees and forces that create hexagonal form.

The overall interplay in the hive as a whole, not in the individual bee, between bee venom, bee food and everything the bees assimilate, on the one hand, and the comb's wax cells, on the other, wonderfully resembles organic processes in the human organism. If we observe bees from the moment they alight on flowers to when they return to the hive, and the products they bring back with them, excrete, and the way they produce their cells, we have before us in this hive activity something that really inwardly closely resembles the action of I, astral body and etheric body in processes inwardly occurring in the brain when a person perceives. Then he takes substances into his powers of perception which work their way into the remarkable structure and shape of the bone cells. In the bee comb we see something that remains soft, unlike the more rigid form assumed by the bone cells; and we see human sense perception figured in the bee sitting on a flower. In fact the whole human organism is encapsulated in what lies between the bee sucking nectar from a flower and the creation of the bee comb.[32]

> Here the silica process is described from a somewhat different perspective. We have sense perception; and Steiner draws our attention to parallels with this in the life of bees—the way they harvest something light, almost insubstantial from blossoms. As in our perception, here something is taken up from the environment, transformed and further processed and assimilated. In bees this gives rise to their hexagonal comb. In relation to our own existence, Steiner describes an astonishing process: substances absorbed by our powers of perception are eventually deposited in our bones.

Silicic acid carries its effects through metabolic pathways into the parts of the human organism where life becomes lifeless. It exists in the blossom, through which formative forces must pass; and it is found in hair, thus where form is outwardly concluded. We meet it in bones in which interior form concludes. It is present too in urine as a product of elimination.

Silica forms the physical foundation of the I organization, which has a shaping, configuring action. This I organization needs the silica process right into the parts of the organism where form and configuration border on the (unconscious) external and internal world. At the organism's periphery, where hair carries silica, the human organization meets and connects with the unconscious outer world. In the bones this organization connects with the unconscious inner world in which the will acts.

The physical foundation for consciousness must develop

in the healthy human organism between the two fields of action of silica. Silica has a twofold task: within us it sets a boundary to mere processes of growth, nutrition, etc. And in an outward direction it separates mere natural effects from the inner organism, so that the latter is not compelled to perpetuate natural actions but, within its own realm, can develop its own.[33]

Overview

> In the lecture cycle on bees, various processes are described which an overview can help us to categorize and understand.
>
> The silica process belongs to etheric life, and is the primary process in the plant. The formic acid process and the carbon dioxide process are governed by the etheric and astral life of animals and human beings.
>
> The silica process is under the sway of the human I organization. It is always close to the mineral realm, to hexagonal, crystalline form, to solidity; and thus it tends also towards the physical and lifeless realms. According to the accounts here, it is central to sugar-formation, which may be connected with the transitional realm, described above, between sweetness and insubstantiality, with the fresh creation of sugars and their ease of assimilation.
>
> The relation of silica to the spiritual domain is presented in the next passage.

This silica represents an infinitely important constituent of our earth. But for someone who, as I suggested, can also understand things spiritually, everything that is embodied in all quartz, in all silica, is at the same time the manifestation of a spiritual element.

Here modern people grow more rebellious than when you talk to them about the spiritual aspect of human beings. They may cope with the latter idea, but if you talk to them about the spirit immanent everywhere in every natural phenomenon or entity they take umbrage. They only want to countenance a nature that is always physical. But that is untrue. Trained spiritual perception shows us that silica, either as quartz crystal or also very finely dispersed silica, is hugely different from carbon dioxide. Nowadays people are accustomed to thinking in terms of known physical attributes: carbon dioxide is carbon and oxygen; silicic acid* is silica [silicon dioxide] and oxygen. Oxygen, silica and carbon are viewed in terms of the properties they show in the test tube, chemical reactions that can be demonstrated in the laboratory.

But there is a spiritual aspect to all this. In fact [...] this silica substance opens a way for the entry of all spirit. Silica substance always allows everything spiritual acting and working in the world to pass through it.

That is the remarkable thing if you look at a quartz crystal: it is like a receiver for the spirit. Twenty-eight per cent of the earth's substance is silica, and spirit passes through it there, just as light passes through something transparent. But quartz also appears in opaque forms, and then is called smoky topaz; and though light does not pass through this, all spirit nevertheless does.

So in nature we meet substances that are simply permeable for the spirit. They are bearers of the spirit. Spirit is in them; they take it up everywhere, at the same time never obstructing its passage. They really are receiver stations for the spirit.[34]

---

* General name for a family of chemical compounds containing the element silicon attached to oxide and hydroxyl groups.

5. HUMAN BEINGS AND BEES

WRITTEN QUESTION: Amongst old peasant beekeepers there is a firm belief that certain soul connections exist between the beekeeper and his charges. It is said that when the beekeeper dies, every hive must be told of it at once. If this is not done, all the colonies will die during the following year. That a certain soul rapport does exist is confirmed by the fact that if you try to work with bees when you're annoyed or irritated you will get stung far more often than if you do the same work in a calm, harmonious frame of mind. Can this belief of old beekeepers actually be based on something real?

Now it would be interesting if Mr Müller would tell us simply whether he thinks such things are groundless. Peasant beekeepers do have the custom, don't they, of telling the bees when someone dies. I'm thinking now of this soul rapport, this relationship between the beekeeper and his bees. Perhaps Mr Müller could say something about this.

Mr Müller tells of two cases in Basel and Zurich that have a bearing on this. In one family the wife—who had worked a lot with the bees—had died, and within a year all the colonies died too. Exactly the same thing happened in Basel. A woman died who had been greatly involved with the bees, of whom there were large stocks. Within a year they were reduced from 28 colonies to six. It is hard to know what the connection is here, or the causes—whether it was to do with the bees or not. Perhaps the bees were sick anyway; or perhaps it was the soul connection.

Let us recall something I once told you about the connection between people and animals. I mentioned it once, or you may also have heard of it. Some time ago there were reports about so-called 'counting horses'. They were asked questions, such as, What is four plus five? Then the questioner counted: one, two, three, four, five, six, seven, eight, nine, and the horse would mark the right answer by stamping with its hoof. These horses could manage fairly complex sums. You may have heard of the fame, especially, of the Elberfeld horses. Investigators went to see what was going on. I myself didn't see the Elberfeld horses, but I did see the 'Counting Horse' belonging to Herr von Osten[35] which had similar arithmetical gifts. How could this be? People were scratching their heads in perplexity. Basically it was rather a terrible thing that horses suddenly started counting. And they did it so fast that they almost outdid a calculator. If this became widespread and horses were taught to count, accountants might soon be worrying about their jobs. Clearly the whole thing was rather alarming.

But scientists, particularly, made a fool of themselves over this whole business. It is easy to see that a horse cannot really count, and so now we must enquire into why it seems to do so, stamping its hoof on the number nine for instance. It is clearly idiotic to think a horse can count. Associate Professor Pfungst, an academic who investigated the phenomenon, also knew this.[36] But he thought up a theory to account for it. Every time Herr von Osten counts, said this learned man, he makes a little bit of a face on the right number, and the horse sees it and stamps at the right moment. And so he watched Herr von Osten's face closely, but he couldn't see anything. But instead of giving up his theory he insisted that the facial expression was so tiny that he couldn't see it, only the horse could. Now from this it follows that a horse can see more than an associate professor.

But he was going up a blind alley. If you're trained in spiritual perception, instead of focusing on a tiny facial expression, you will find that things were like this. Here we have the horse, and there we have Herr von Osten, holding the horse's reins. And in his right pocket he had lots of little sugar lumps. Herr von Osten kept giving the horse sugar. The horse was delighted and loved Herr von Osten very much. This love kept growing as the sugar kept coming, creating a warm and heartfelt relationship between the two of them. Herr von Osten did not need to wrinkle his face but only needed to think 'the right answer is nine' and the horse sensed it since animals have a far subtler sense of what's going on around them than we do. They sense what's happening in someone's head even if unaccompanied by a facial expression that a horse could see but a person couldn't. The horse feels what's happening in the brain of someone when he thinks 'nine'. And the horse stamps. If the horse had not received the sugar, its love would have turned a little into hatred instead, and it wouldn't have stamped any more.

So you see that the animal has a fine, subtle sense for things—not for small facial expressions but for things that are actually invisible. You have to observe such things closely and then you discover how animals have a wonderful sensitivity.

And now imagine there's a swarm of bees around you and you're frightened. Well, the bees feel this, it's undeniable. What happens when you're frightened? As you know, you grow pale, and the blood drains away from the surface of the skin. When a bee approaches someone who feels fear, she senses this hexagonal, crystallizing force more strongly in the person than at other times when the blood is flowing through the skin's surface, and she tries to get the honey or wax from him. By contrast, when he does things equably, and his blood flows evenly, the bee notices something quite different—that his blood possesses the same hexagonal, crystallizing force [as she does].

If a person is angry, and deals with bees in an angry state—yes, anger makes you red, the blood flows strongly; the blood wants to absorb hexagonal forces. The bee, with her fine sensitivity, notices this and thinks you want to take this force away from her, and so she stings you. Fine sensitivities to natural forces are at work here, are in play.

And then comes habituation. A beekeeper doesn't approach the hive in the same way as anyone else. If I can put it like this, the bees feel his whole emanation, they sense his qualities, and they get used to this. If he dies they have to adjust, and this is a major thing for them. Even amongst dogs it has been known for a dog to pine away and die after his master's death because he couldn't get used to a new master. So why should it surprise us if bees, who possess such a subtle chemical sensitivity, become so accustomed to the beekeeper that they cannot immediately adjust to a new one? These are certainly real factors.

But you might ask whether dogs and horses are not very different in kind from these tiny insects. Well, I don't know if you've encountered this yourselves, but there are people who have what is known as 'green fingers'. When they grow plants everything thrives, while others may tend their garden with equal care but nothing much comes of it. This is due to the emanation a person has, which in one case acts favourably on flowers and in another does not. Some people cannot grow flowers at all. And this negative influence acts chiefly on the force in the flower that creates nectar, that develops sweetness. So we can say that a human being even exerts an influence on flowers, and pre-eminently on bees.

Instead of being astonished at this, we should just try to hold the facts together in our mind, and then we will see how such things can be. And we can also take account of them in practice.

Second question: An old farmer's saying tells us that if it rains on 3 May, the Feast of the Cross, this washes nectar from the flowers and trees, and so there will be no honey yield that year. My observations over the past four years seem to confirm that there is some truth in this. Is such a thing possible?

Yes, this is something that leads us very deep into natural influences. The precise date of 3 May, the Feast of the Cross, is of less significance. But this particular season, this period, is important. What does it mean if it rains around then, at the beginning of May? I once told you that at present the beginning of spring falls in the sign of Pisces, and the sun remains in this sign until around 23 April, then it moves on into Aries. So at the beginning of May the sun's rays come from a quite different corner of the cosmos than at other times. Let's assume the weather is fine at the beginning of May, and thus on 3 May. What does this mean? It means this. On 3 May the sun has a strong power over everything terrestrial. Everything that happens on the earth is subject to the power of the sun when the weather is fine. So if it rains at the beginning of May, on 3 May, this means that the earth's power is stronger and dispels the sun's action.

And all this has a huge significance for plant growth. When the sun's power can act strongly in the sunshine coming from the region of Aries, can exert its whole vigour on the flowers, they develop this sweet substance of nectar and it is available to the bees. But when the earth has the upper hand because it's raining at the time, flowers cannot develop under the influence of sun rays shining upon them from the direction of Aries. So they have to wait for a later date, or what has so far developed may be interrupted altogether. Then they don't form nectar properly, and the bees find none.

A thing like this is understandable if we know that everything that happens on the earth, as I have often said, is influenced by the cosmos, by what is beyond the earth. Rain means that the power of the sun is dispelled, whereas fine weather means the sun can unfold its full power. It is not just a matter of solar power in general but that its power comes to us from a particular region of the cosmos, in this case from Aries. The sun's rays bring a different quality from each part of the cosmos. It is not the sun alone, but the fact that the sun absorbs what Aries gives it, and passes this on in the sunshine. The quality is quite different depending on whether the sun is shining on the earth at the beginning of May or the end of May. At the beginning of May, the full strength of Aries is still at work. By the end of May, the force of Taurus comes into play, and this brings something that can no longer act on plants with the same vigour, but instead hardens and dries them; and in turn this means that the plants can no longer develop their nectar-forming activity.

Thus the old farmers' sayings have their good reasons and we should attend to them. Of course, as I've said in the past, consciousness of such things has been lost, and instead superstitions hold sway. Of course, when we no longer discern realities, we can succumb to superstitions. Then old proverbs may be of no more value than saying something like, 'If the cock crows on the dunghill, the weather will change or stay as it is.' This is not true of all these proverbs, though;

some of them are founded on deep wisdom, and then you have to enquire carefully into what underlies them. The farmers who rely on them sometimes do very well with them! Deeper insights can lead us to reviving some of these farmer's proverbs and making use of them.[37]

Mr Erbsmehl puts forward the view that in modern beekeeping profitability is the prime focus. Thus it is concerned with material things only. The Swiss Beekeeper's Journal, *issue 10, October 1923, states: 'Honey is very largely a luxury item, and those who buy it can pay the proper price.'*[38] *Later in the same issue*[39] *the tale is recounted of a certain man, Baldensberger by name, who travelled through Spain and met a beekeeper and a number of blossoming children. Asked where he sold his honey, he replied, 'Here are my customers.' Here in Europe people try to squeeze a large profit from their honey. An employer with many employees will focus on squeezing as much as possible from his produce, and the same applies to beekeepers and their bees.*

Then he asks also if there is anything in the claim that moonlight has a certain effect on nectar formation in flowers (issue 11 of the same journal).[40]

Mr Müller replies that Mr Erbsmehl can see from the journal that the beekeeper who doesn't sell his honey [but gives it to his children] only operates on a small scale. Erbsmehl, he says, is ignoring the reality of beekeeping today, and everything involved in it, including costs. If commercial viability is not considered, as in other businesses, then one would have to give up beekeeping. And if honey wasn't produced in this way, through artificial breeding, there would be much less of it around. Sometimes there are only one or two kilos of honey, occasionally a little fir honey too,[41] *that one can take from a hive if the colony is to stay healthy. That is all. If it's a bad year after this, they won't have enough to last until April or May. Colonies have to be helped to overwinter with artificial feed—sugar, camomile tea, thyme and a little salt. In a modern operation, hours of work are carefully noted—how much time the beekeeper has taken; five and a half hours labour is calculated to cost 1 franc, or 1.50. Thus [a jar of] honey is valued at 7 francs. And then you also have to allow for depreciation: the frames or something else need replacing. The whole business has to break even. If beekeepers don't move with the times they will get nowhere. Mr Erbsmehl can do it, but if I have large stocks I have to count up the sums and be aware of a shortfall if I sell the honey for 6 francs. American beekeepers hold exactly the same view as this.*

He goes on to say that he cannot understand that colonies might die off after 80 to 100 years. He cannot comprehend what Doctor Steiner means by saying that in 50 to 100 years artificial breeding will lead to this state of affairs.

As regards the second point, telling the bees about the death of the beekeeper, he had already mentioned an instance where the majority of bees died after the death of their beekeeper; he cannot himself understand how this can be.

As regards artificial honey in hotels, syrup, he would like to say that first-class hotels largely sell American bee honey. When bees are fed this, they perish. And yet it is still a bee product.

And as to bee stings: sweat is the worst thing for bees. If you hear piping or vibrating tones in the hive, it is best to stay away.

As to the question of how a bee sting can affect someone, he knows of an instance that he wishes to relate. A vigorous man, even stronger than Mr Binder, was stung by a bee. He cried out, 'Help, I've been stung!' He was extraordinarily sensitive to stings, he also had a heart complaint. Perhaps Doctor Steiner can tell us how dangerous a bee sting might actually be. For example, it is said that three stings from a hornet can kill a horse. I found a hornet's nest recently amongst my bees. I took away the brood. The hornets were so cowardly that they did not sting in the dark; in the open air they might have done.

To speak as rationally as possible about these things, let's first consider how bees recognize the beekeeper.

You made a judgement which is of course entirely justified in rational terms. But now let me say this. Imagine that you have a friend whom you get to know, say, in 1915. This friend stays here in Europe while you go to America, only returning in 1925. Let's say your friend lives in Arlesheim. So you come back and visit him there, and recognize him. But what has occurred in the intervening period? As we have discussed before, the substance, the matter composing the human body is completely renewed in seven or eight years. There is nothing left of the substance that was there before. In other words, when you see your friend again after ten years, there is nothing of him, really nothing at all that you saw ten years before when you left. And nevertheless you recognize him. If we look at him outwardly, he looks like a coherent whole. But if you were to examine this friend of yours through a large enough magnifying glass you would see the blood vessels in his head. But now imagine a very high magnification: this shows blood no longer to be a fluid only but to consist of many small dots. In fact they look like little creatures—and they are not at rest but quivering the whole time. In fact, looking at this you would say that it bears a remarkable similarity to a swarm of bees.

At a large enough magnification, a person's substance looks just like a swarm of bees. Understanding this, it must seem incomprehensible that we recognize someone again after ten years, since not a single one of all these tiny moving dots are the same as before. His eyes too are composed of completely different dots—these small creatures are there too, and yet we do recognize someone.

You see, it is absolutely unnecessary for our recognition to be based on all these individual animal-like and plantlike entities of which we are made. No, someone recognizes us as a whole person; and likewise the colony is not so many thousands of separate bees but it's a whole, a whole entity or being. It is the whole colony that recognizes us again.

If we had a miniaturizing glass instead of a magnifying glass, you could bring all these bees together, resolve them into one entity, like a human muscle. So this is how we must consider the bees: not as separate bees but as a whole entity that absolutely belongs together. Mere reason cannot grasp this. One needs the ability to see the overall whole. That is why the life of bee colonies is so very instructive; the hive completely refutes rational considerations. Our deliberations always tell us that what we observe ought to be different. But in the hive the most wonderful things occur, and not as we might initially think when we employ reason. A change such as the death of the beekeeper does have a certain effect on the hive, and it cannot be denied. It exists and experience shows us this is so. If you have seen many hives and beekeepers you do perceive this.

I can tell you that when I was a lad I came into very close contact with beekeeping in all kinds of ways. I was keenly interested in it at the time. Back then I was not so concerned as I later was with financial and economic considerations relating to beekeeping. At the time, honey was so expensive and my parents so poor that we could not afford it. But our neighbours always gave us some, usually at Christmas; and we were given so much honey the rest of the year too that we always had supplies of it. People gave it away; and so I wasn't particularly concerned in my boyhood with the financial aspects of beekeeping since I ate a great deal of honey we had been given. How could this be done? Nowadays it is much less likely that honey will be given away free. But back then, in my neighbourhood, beekeepers were mostly farmers too, and regarded beekeeping as part and parcel of agriculture.

This is very different from someone setting up business as a beekeeper, and having to live from his enterprise. Beekeeping can be included in farming and is scarcely noticed. The labour involved in it doesn't figure, it's just something done at odd moments. When part of farming, it's a job done when you have spare time, or when you shift another job to another time. At any rate, as I experienced it, honey was harvested in between other things, and we had a sense of honey being so valuable that you couldn't buy it. In fact, in a sense, this is absolutely right, and today everything is subject to mistaken price relationships. Price is something that needs far more wide-ranging economic discussions. Nothing much comes of discussing the price of single foods and items; and honey is a food, not just a special delicacy or a luxury item. In a healthy social order, a healthy price for honey would quite naturally come about, there's no doubt about this.

But we definitely do not live in healthy social conditions today, and so all questions are distorted. For instance, let's consider large agricultural enterprises. Usually what the farm's business manager says—not actually a farmer at all—about the quantity of milk he gets from his cows is dire. He gets so many litres a day from his cows that anyone who knows something about dairy cattle will say: 'That's impossible!' And yet it happens. Those amounts of milk are obtained from cows. Sometimes they get more or less double the amount of milk that a cow can realistically produce. This makes the estate very profitable of course. Nor can one even say that the milk is noticeably poorer in quality than milk obtained under natural circumstances. So you can't initially prove that anything bad is happening here.

But let me give you the following example. We have been doing trials on a remedy against foot-and-mouth disease in cattle, one among many trials we have done in recent years.[42] We undertook these trials on large estates but also on small farms where cows do not have to give so much milk as in big enterprises. We learned a good deal in the process, while testing the remedy for foot-and-mouth. The trials were not completed because officialdom didn't want to pursue it, and because all kinds of compromises are needed nowadays. But the remedy was extremely effective. And in a somewhat modified form it is being used to good effect, for example, to combat distemper in dogs.

If you undertake such trials you find the following. You discover that calves from cows required to give excessive amounts of milk are considerably weaker than others. You can tell this from the action of the remedy: the effect or non-effect is hugely magnified. If the calf survives foot-and-mouth, it does grow and develop, but if it comes from a cow that is overfed and thus forced to give too much milk, it will be weaker than calves from cows that are not compelled to give so much milk. You can see this in the first, second, third and fourth generation. The difference is so small as to be scarcely noticed for now. Large-scale milk production businesses have only existed for a short while; but it is quite clear to me that if we continue maltreating cows in this way, making them produce over 30 litres of milk a day, then eventually the whole business will collapse completely. It will be unavoidable.

Naturally things are not as bad as that with artificial beekeeping methods since the bee is a creature that continually helps itself, is much closer to nature than cows bred and managed as I described. It is not even so bad to maltreat cows like this, breed them for high milk yields, as long as you still let them graze in the fields. But that is changing now too: these big outfits keep the cattle inside, in barns, thus removing a cow completely from her natural environment.

That cannot be done of course in beekeeping, since by their very nature bees remain connected with their surroundings. They help themselves and recover. And this self-help is something very wonderful in the hive. We come now to what Mr Müller said about hornets, which he sometimes finds amongst his hives. He says they didn't sting him, although in other circumstances it is a good idea to keep a safe distance from a hornet.

Here I'd like to say something else. Beekeepers will know that it is sometimes necessary to empty a hive. On one occasion I looked into an empty hive and saw something remarkable there—a sort of lump. I didn't at first know what it was. For some unknown reason the bees had created a lump there, using the substances with which they make their other products. There it was, a lump made of all kinds of resins, glues, lime-type substance—which the bees also gather—and wax, etc. I was curious to know what this was, and opening it up I found a dead mouse inside.

The mouse had crawled into the hive and had died there. You can imagine how unpleasant the smell of the mouse corpse must have been for the bees! In this unusual instance, the whole hive instinctively encased it. When I opened up this casing, there was a horrible stink; but while enclosed, the smell stayed inside the shell of wax and propolis. So the instinct of the whole hive extends not only to building comb, feeding the larvae, etc. but also to doing something unusual, for instance if a mouse crawls in. Since the bees were unable to drag the mouse out, they helped themselves by forming a shell around it. Others have told me that they do the same with slugs that find their way into the hive. The hive possesses not only the usual instincts therefore, but real healing instincts. These are very much alive in the hive.

Now if there is a hornet's nest in the hive, the bees do not cover it in the same way with a solid casing, but they do continually surround the nest with their sting secretions, and in consequence the hornets lose the energy, the strength to attack. Just as the smell of the dead mouse can no longer emanate through the hive, the hornet—even though not firmly encased—lives continually in the vapours with which the bees surround it, weakening it so that it cannot any longer harm them. The hornets completely lose the strength and energy to defend themselves when the beekeeper comes near them.

It is true to say that we only comprehend bees if we go beyond mere intellect and observe their life with a kind of inner vision. The picture they give us is a wonderful one: the hive is a whole entity and must be understood as such. But in terms of a whole organism, damage does not immediately become apparent.

If we have a good knowledge of human nature we can observe people—they certainly exist—who are still fairly energetic at the age of 65 or 66. Someone else at the same age is less vigorous, suffering perhaps from arteriosclerosis and suchlike. It is extremely interesting to see this in connection with what occurred in a person's childhood.

For instance, you can give a child milk from cows that are

overfed with fodder grown on a chalky soil. In consequence the child absorbs something of the lime in the soil. This may not become apparent immediately. A doctor of the modern type will show you a child of both types, one raised on cow's milk with a lime constituent and another raised on mother's milk, and will say, look, this makes no difference. But the child reared on mother's milk will still be fresh and energetic at 65 or 66, while the child raised on cow's milk has arteriosclerosis by that age. This is because the human being is a whole, and so the influences active in one period work on into a much later one. Something can cause no ill effects in one period and yet it goes on acting. And that's what I mean when I say that we cannot yet discern what the effects of artificial beekeeping methods are; we will have to wait and see in 50, 60 or even 100 years from now. It is quite understandable if someone says he doesn't see why anything should change in 50, 60 or 100 years' time, and does not wish to accept such an idea. I once encountered the same thing on an estate. Believe me, the farmer wanted to throttle me when I started to say that a farm ought not to produce so much milk, for cattle rearing will go downhill much faster; at this rate, it will be ruined in a quarter of a century.

You cannot say anything against artificial beekeeping methods today since we live in circumstances where there is no room at all for social innovation. But it must be recognized that it makes a difference whether you give nature free rein and merely try to guide it gently or if you introduce something artificial into the process. But I certainly do not wish to take issue with what Mr Müller has said. It is quite true that at present these things cannot be discerned, and we will have to defer judgement on the matter. Let's speak again in a hundred years' time, Mr Müller, and then see if your views have changed or not. These are things that cannot be decided at present.

Mr Erbsmehl again makes the point that everything in modern beekeeping is geared to profitability.

The more someone practises beekeeping alongside other work, the more you will find him to share the view of the Spaniard you spoke of. In general this may no longer be the case, but 50 or 60 years ago the farmer didn't worry much about the yield he got from his bees. This was always something he left out of his accounts. He either gave the honey away or, if he sold it, he gave the money he made from it to his children for their piggy-banks or similar. Nowadays circumstances have changed radically. It is unimaginable now for someone who works for an hourly wage not to consider profitability. He is compelled to through the whole circumstances in which he lives.

Beekeepers today, who work in other jobs too, have to take a break from this other work in order to undertake beekeeping properly. Isn't that so? ('Definitely.') Well, then of course, they work out the earnings lost from their other work.

Consider this: beekeeping is such an ancient custom that there is no longer any outward evidence to say what beekeeping was like when bees were still completely wild. People nowadays only know our bees, European honey bees. They only know domesticated beekeeping. As far as I am aware, natural histories give the name of the bee widespread in Europe as the 'common domestic bee'. So we are only familiar now with bees as domesticated creatures. Let us take special note of this. We are not aware how things were in an ancient, natural condition. Beekeeping is very ancient, and in such ancient concerns the price arises from quite different considerations than those that usually apply today. Mostly people work in pursuits whose origin we can very easily trace. So there are two approaches to setting prices. In the case of the bees, this practice goes back to time immemorial, and so prices do not apply in the same way as in the metal or timber industries, which are only a few decades old. Only when we have healthy social conditions will it become apparent how the price of honey should be determined, and how this price once arose from quite different circumstances.

People have no idea how difficult it is to speak of pricing nowadays, for to do so requires profound insight into actual circumstances. Recently I had a really wonderful experience in relation to pricing, and I'd like to tell you about it since it is very interesting.

A university professor whom I know wrote a book on economics. He gave me a copy of the book when I was on a lecture tour. I told him I'd leaf through it—I couldn't read the whole thing straight away—when I arrived in a particular place where I'd be staying for two or three days. Well, I looked through the book and I said this to him: I was pleased to see that you have an index in your book—this doesn't always please me, but I was glad to see your book had an index. But when I looked up the word 'price' in the index, there was no entry for it, absolutely none!

The man had written a book on economics, with absolutely no mention of pricing! This is rather typical. Economists today are unable even to perceive the most important problem in economics. They leave out pricing, and yet pricing is all-important. They just don't recognize it. My experience with this book was a clear illustration of this.

Here too we must hope people will realize that healthier social conditions must gradually be introduced. Then, I believe, there will be less talk of profitability or unprofitability. These are terms that relate to competition and competitiveness, albeit not competition with other producers of the same goods, but with producers of different goods.

If you looked back with me to my youth and considered beekeeping in the region where I lived, where beekeeping was something only done by farmers, you would find that these farmers were stout and sturdy people. I can't even say they resembled one among your company, for no one here is as fat as they were. The price of honey was such that no one would have set up business as a beekeeper alone, selling his honey and living from the proceeds. If there had been such a beekeeper and he had been standing next to a farmer, they would have looked like this [see drawing]. Here's the farmer, and there's the beekeeper. That would never have worked! The farmer simply didn't reckon on proceeds from the bees, whereas the beekeeper would have had to rely on them. So it wouldn't have worked. As soon as we start to talk about profitability, we have to have a thorough understanding of economics, and take account of it.

Now I'll answer a couple of questions that arose in relation to what was said above.

Question: There are people who cannot eat honey at all, and immediately get stomach problems if they do. Is there any remedy that can keep this in check?

People who can't digest honey usually have a premature tendency to sclerosis, hardening of the whole body, so that bodily metabolism is very slow. Because honey speeds up the metabolism, they can't cope with it. Their metabolism is too slow; the honey tries to make it quicker, and they end up in conflict with their own metabolism and get stomach problems that express themselves in a whole range of disorders. Everyone should really be able to enjoy a little honey—I mean not only enjoy it, but should have the capacity to digest it.

If you have someone who is intolerant of honey, you should first of all determine where the causes lie. There isn't a general remedy as such, but depending on what is causing hardening, sclerosis, you will have to cure it with different remedies. The following case might arise for instance.

Let us say someone is intolerant of honey and gets stomach problems. Then we must ask whether he tends to sclerosis of the head, of the veins, vessels and arteries in the head. In this case it may be that he becomes intolerant of honey at a certain age. Then you will need a phosphor preparation. Given this cure, he will begin to be able to digest honey again.

Or the problem might lie in the lungs. Then instead of phosphor you should give a sulphur remedy. That's my answer to this question. It is not a matter of generally determining that someone gets stomach problems when he eats honey. Instead, if someone becomes intolerant of honey at a certain age, this is a disorder. A healthy person can digest honey. If someone can't, he is ill, and then you must seek the specific illness and try to cure it.

Intolerance of honey is not of course as bad as intolerance of sugar, diabetes mellitus. That is of course a graver condition, and likewise needs medical attention.

Question: Like most insects, bees always fly towards the light of a candle or lantern in the dark. Experienced beekeepers have often told me that bees react much less to electric light. If you approach the bees with an electric torch, they remain quite calm, as if they do not perceive the light at all. They only grow restless after a while. A paraffin lamp or candlelight elicits a far quicker and stronger response. Is there an explanation for this?

Mr Müller says he has noticed the same thing.

Well, in the old Goetheanum building you will have seen that the cupolas were painted in various colours. These paints were made from pure plant substances. Because of their composition, if sunlight had shone into the building the colours would eventually have faded altogether, perhaps in months or years, so that you would no longer have seen what had been painted there. But exposed to electric light, they would have been preserved. Sunlight would have faded them altogether but under electric light they were preserved.

So you can see that sunlight, which exerts chemical effects—and as you said, the bees notice this—works very differently from electric light. Electric light exerts a much more hardening effect on all substance, does not dissolve it in the same way. In consequence the bee upon whom electric light shines suffers something like a very mild catalepsy, which is not the case in sunlight. And then of course it recovers again.

Question: As regards the influence of the zodiac signs on honey production, in rural districts people still place great value on sowing their seed when, say, the moon is in Gemini and so on. Is there good reason for this or are they simply viewing each zodiac sign in terms of superficial characteristics?

Of course no one has ever properly investigated these matters scientifically. They can be investigated. The hive as a whole is affected by the influences I described before. In certain ways the bees, and the queen specifically, are solar creatures. The sun's passage through the zodiac has the greatest influence on bees. But the bees are also of course dependent on what they find in the plants; and here it is certainly true to say that sowing, the scattering of seed, can have a great deal to do with the passage of the moon through the signs of the zodiac, and this in turn affects the substances that become available to bees in the flowers. Such things are not random by any means. But as usually described, these things are presented fairly superficially. They ought to be properly scientifically investigated.

We have run out of time for today. We'll meet again on Saturday at 9 to take these things further.[43] It seems to me that many of you still have questions. Beekeeping is so fine and useful an activity that one can never have too many questions. Do ask questions of each other too, and ask Mr Müller and me. I imagine we will slowly come to a gentle balance of contrasting views. There is no need to start stinging too quickly like the bees themselves; we can come to gentle agreement. But we should be forthright with our questions, not hold back.[44]

With their question about the soul connection between beekeeper and bees, the workers pick up on a theme that is still as topical today. As mentioned in Chapter 1, such a relationship is only conceivable if we assume a bee colony is a single, whole organism, a soul-spiritual unity. The being of the colony is the 'opposite number' with whom a communicating relationship and soul connection arises. We can encounter the colony at two levels of communication: firstly the whole nature of bees in general, founded on an encounter with the archetype or being of the bee. Here we seek to understand the nature of a colony in relation to the human being. This is, really, the aim of this cycle of lectures. It is an ongoing engagement with bees leading to a new, living relationship with them in which we acknowledge this group soul as such. But secondly, also, this question concerns our individual relationship to individual colonies, arising through practical work with them and close observation.

Rudolf Steiner characterizes three levels of relationship between the human being and animal. In his story about Herr Osten's horse, we see intentional training of an animal that leaves its traces in the creature's subtle sensitivity. Though the bees cannot be trained in this way, we can recognize their alertness to the beekeeper, and, as in the example of the horse, an intense encounter is at work. Secondly he speaks of gardeners with 'green fingers' also in connection with the bees. Beekeepers are familiar with this phenomenon: some people always or mostly have fine, healthy colonies, while for others nothing really succeeds whatever they do. Thirdly he refers to the death of a beekeeper. That colonies die after the death of their beekeeper is not something we hear of any more today This must have involved a deep, instinctive connection which, even in those days, was only present still in old rural communities.

Today, given prevailing conditions, bees make big demands on those who would like to keep hives. It is no longer enough to feel drawn to keeping bees, to set up a hive in the garden and leave them to it. Colonies can no longer survive unaided but require continual supervision, at least during the six-month summer season. You have to observe them closely and attentively and take the right measures in due time. Demands on beekeepers have grown.

The death of bees, of whole colonies—colony collapse as it's called—is a burning issue today. The phenomenon has diverse causes: lack of food sources, the varroa mite, modern agriculture, and modern beekeeping methods themselves.

Bees naturally form large colonies from April to September, to make use of a profusion of blossoms and to store up large supplies within a short period. But the reality of our cultivated landscapes now is that a wealth of flowers appear only in spring; after this the bees can no longer find enough food sources and are compelled to consume their stores. In modern agriculture, meadows are used early before plants blossom, and flowering weeds are treated with herbicides. From mid-May onwards there are scarcely any blossoming plants left. Previously ubiquitous summer plant cover such as sainfoin or clover has disappeared as modern agriculture progressed; loss of hedgerows and untilled areas has led to the disappearance of flowering bushes like buckthorn. If woods and copses are uprooted too, the bees may be starving by July. Lack of flowers for the summer subsistence of bees turns a landscape, really, into a green desert.

Another major and direct cause of colony collapse is the varroa mite—a parasitic mite originating in Asia, which became widespread last century and attacks both brood and bees themselves. The mites can only be kept in check by treating the bees twice a year: in summer with formic acid and in winter with oxalic acid. This reduces their numbers enough to enable colonies to survive. The mite's association with viruses has increased its risk over the last ten years. Mite infestation has to be carefully monitored, since delayed treatment in summer can lead to the loss of a whole colony the following winter.

Modern agriculture repeatedly burdens the land with a toxic mix of herbicides and pesticides. While bees do not usually die directly from these substances, 'sublethal' doses of them accumulate slowly and weaken their capacity to survive.

Beekeeping methods themselves are also problematic. Trials have shown that single hives have remarkable regenerative capacities but that many colonies living side by side in the same place restrict these self-healing powers since flying bees keep spreading disease.

This brief survey of the problem shows how intensive agriculture, along with the spread of disease through global trade in bee queens, and even through beekeepers themselves, is hampering bees' survival. These human-instigated problems require much effort to remedy.

The issue of remedy and rebalancing through a new approach to beekeeping is not new, but has certainly grown much more topical since Steiner's day. It will form the theme of the next chapter.

6. SUPPORTING THE BEES

Mr Dollinger asks about comb. He says there are people who eat the comb along with honey, and in inns and taverns it used to be put on the table sometimes. He'd like to know if there is any harm in eating it.

As regards bee diseases, he thinks that there used not to be so many of them as today, now that bees are more commercially exploited.

Mr Müller says that eating comb is a personal taste—and of course this refers to natural not artificial comb.

As to bee diseases, he says, they are not due to commercial exploitation but were simply disregarded more in the past. Colonies used to be larger and so less notice was taken of such problems. A new bee disease has also arrived in Switzerland from England.

Mr Erbsmehl thinks the problem might be related to artificial fertilizer, which can make plants more susceptible to disease.

It is right to say that eating comb along with honey is a personal preference, a particular sort of delicacy. In such things of course we have to see how it agrees with a person, a question that can only of course be answered in medical terms. We can only say something about this if we observe the state of health of people who eat the comb, that is wax. Though I have witnessed people who ate comb with honey, they would always spit out the wax again after sucking the honey from it. I haven't been able to study those who ate a large quantity of wax with the honey.

People vary in what agrees with them. There could be people who develop a stomach disorder from eating wax, and then they should refrain from it. But some might be able to digest the wax without a problem, excreting any residues. In their case one could say that by eating wax with honey, thus keeping the honey with the wax for as long as possible, the honey is digested more in the intestinal tract whereas otherwise it would only be digested later, after it leaves the intestine, in the lymph vessels and so forth. In turn this is connected with a person's whole state of health. Some digest more through the intestine, others through the lymph vessels. You can't say one is better, the other worse; it just depends on the individual. You could only find more evidence either way by setting up a trial in which you let one group eat honey with the comb and another honey without, and then investigated how these two things relate to each other.

As regards bee diseases, what applies to illnesses in general applies here too, and we must note what Mr Müller says. It is likewise true of human beings that people used to give little attention to certain conditions, whereas now they are the subject of close study. They were disregarded in the past, as you say, and nowadays they figure large.

But there's another important factor at work here too. In the old days a beekeeper had a very instinctive approach, doing much that he could not necessarily explain. Humanity has lost such instincts. Today everyone wants to know why things are done as they are. But to determine this, we have to study things closely and thoroughly—something that is as yet difficult with the level of knowledge people have. In the old days a beekeeper had a strong instinct for handling his bees in a really intimate, personal way. And now consider this: there is a difference in the fact that beekeepers used to use straw skeps but now use wooden boxes. Wood is a different material from what used to be used, woven straw and suchlike. Straw draws from the atmosphere quite different substances than abstract wood. So even in this outward circumstance we find a difference of approach.

In summary, the beekeeper used to do things instinctively, often not exactly knowing why he did things—why, for instance, he placed his hives in a particular position relative to the prevailing wind, and suchlike. Today hives are placed wherever it seems useful, wherever there is room and so on. Climatic conditions are still considered, but less than they used to be.

Mr Müller says that he pays a lot of attention to this. He keeps his hives on a ridge protected from the north wind, and not too exposed to the east wind either.

Wood is less sensitive for such things than straw. But I'm not starting a campaign to go back to straw skeps. It is just that such things are real and have a strong effect, and no doubt play into the life of the bees and the way they do all their work in the hive. A huge amount is going on inside their bodies when they harvest the nectar and convert it into honey. This is a very great deal of work for them. How do bees accomplish it? By virtue of a very specific relationship between two juices inside the bee, its stomach juices and its blood. If you study a bee you find it contains white gastric juice and somewhat reddish blood. These largely constitute it. All its other parts are as it were organized through the influences of these two juices.

In this relationship between gastric juice and blood, the two substances are significantly different from each other. The gastric juice is acidic while the blood is alkaline—in other words, non-acidic in chemical terms (it can only be somewhat acidified). Now if the gastric juice does not have sufficient acidity, something immediately happens in the bee that disturbs her inner organism while she is making honey. And the blood is only sufficiently strengthened again through the presence of the right climatic conditions, light conditions, warmth conditions and so on.

If we are to properly combat the new bee diseases that have arisen, it will be very important also to have the right influence on the relationship between gastric juice and blood in the bee. Since beekeeping can no longer be conducted in the same simple way it once was, we won't achieve this through climatic and warmth conditions. These no longer work so strongly on the newer types of hive. It will be necessary instead to enquire into what acts best on the blood of the bee, and for beekeepers to try to ensure the right blood constitution. This will depend on the following.

As you know, in some years the bees have to gather their honey almost exclusively from trees, and in such seasons the composition of their blood is at great risk. They become sick more easily than in other years. In future, therefore, a beekeeper will need to set up a kind of small greenhouse—no more than a small one—in which he cultivates plants that the bees love and need at a particular time of year, so that one has at least a small flower bed that the bees can come to in May, for example. They will find their own way to it if it is there, when the plants they usually need are not thriving in that particular May, or are not there at all.

So by cultivating plants near the hive to support the bees, it is certain you can remedy such diseases. That is the kind of thing I would recommend. They are just suggestions but I'm sure they will prove effective since they are drawn from knowledge of the nature of the bee. If a beekeeper will try them, they will have good results. You will find this will combat bee diseases. But to remain practical we have to consider all circumstances.

This is something I am not claiming definitively, but am only saying that it becomes apparent from the whole nature of the bee. We can try cultivating plants that have done badly at a particular season, that have not appeared, and we will probably greatly benefit the health of the bees by doing so. I am convinced such things will work, and that we can discover them when we really closely engage with bees and their nature. Nowadays it isn't possible simply to return to the past. Just as little as one need be politically reactionary does one need to be reactionary in other realms. The world progresses and you have to go with it. But as old ways are left behind, this must in turn be balanced by something else that returns things to health.

Mr Müller says that beekeepers are already working towards flower cultivation to support the bees. For instance, large swathes of yellow crocuses have been planted specifically to provide blossoms for bees, likewise other plants with similar small, yellow flowers and so forth. American clover is also being widely planted; it grows to a height of about two metres, and blossoms through the year. It is only cut down in the autumn, and before that left to blossom for the bees.

Certainly, a beginning has been made with such things, but as yet too little is known about the interrelationships. What you first said is a good idea, and this can be continued.

What you said about American clover which blossoms through the year—well, people will stop doing that since it doesn't enhance the bees' blood; it just stimulates them or spurs them on for a short while. This American clover is like trying to cure someone with alcohol—it stimulates the bees for a little while only. We should take the greatest care not to introduce alien substances to the bees, since by their very nature they are connected with a particular region. You can tell this from the fact that bees from other places have a different appearance. We have this central European bee here, the common domestic bee I spoke of before. The Italian bee looks very different, the Slovenian bee different again.[45] Bees are very much accustomed to their own regions, and they gain no long-term benefit at all from being given honey that comes from faraway locations. This creates serious problems for their body, things start rumbling as they try to adapt to how things would be if they lived where the clover flowers come from [from which the honey was made]. It will soon become apparent that this works only for a couple of years, but after that you'll have a lot of bother. You say, quite correctly, there are no clear guidelines yet. But these things will emerge in time, and the proof of the pudding will be in the eating—either people will continue with them because they work, or will give them up, in the same way as happened with wine. You may remember that in the 1870s and '80s vines in many areas of Europe were suddenly attacked by the vine weevil. At the time I was able to study this closely as I had a good friend who was a farmer and winegrower.[46] He also published a farmers' journal and concerned himself with this question. People wondered why the American vines hadn't yet succumbed to the weevils, seemed insusceptible to it. But what came out is that the means used to fight the weevils on American vines could not be used for European vines. And this meant that when people began to plant American vines in Europe it was possible to keep the latter healthy while the European ones succumbed to infestation. They had to give up the European vines altogether and Americanize all viticulture. And then the whole nature of viticulture changes, and becomes something different. That is what happened in many regions.

One cannot think so mechanically, but one has to realize that a living plant or creature is specifically adapted to its own locality. You have to take account of this—otherwise you may have short-lived success, but not in the longer term.

Are there any other questions, or would you gentlemen prefer to get down to eating honey rather than discussing it at such length? Maybe one or other of you still has a question?

I'd like to return briefly to what is really at work when the bees make honey. You see, this is something very wonderful—that these little fellows are able to transform what they suck from flowers and blossoms into this extraordinarily healthy substance of honey, which could play a far more important part in human nutrition than it does today. People ought to recognize how hugely important it is to eat honey. If it were possible to have more say in the whole field of what I will call 'social medicine', it would be extremely beneficial to get engaged couples to eat honey as preparation for having children. Their children would not get rickets, since honey, as it is further digested and assimilated, is able to act on reproductive capacities and endow children with the right kind of form and constitution. When parents, especially the mother, eat honey, this will act on their children's bone structure.

When deeper insight into such things develops, they will become a matter of course. Instead of all the poppycock you find in scientific journals today, people will come to know what they should eat at different ages and stages of life. This will be extremely beneficial, and will greatly enhance human health, and human energy and strength particularly. People at present give very little credence to such things. Today someone is simply pleased if his children don't get rickets, but he doesn't think further about it—he just accepts this state of affairs. Only someone who does have a child with rickets will worry or complain about it. It is really high time to introduce social-medical measures in order to create a situation that people ought to regard as normal.

There needs to be recognition of the enormous benefits gained by working in this way; it seems to me that if people were to realize that we arrive at such ideas through spiritual science, they would turn in far larger numbers to the spirit than they do at present, when they are told only that praying will suffice. Things such as I have spoken of here are perceived through the spirit. Science fails to acknowledge them but they are things that can be discovered—for instance the great benefit of eating honey when you're engaged to be married.[47]

The lecture of 10 December 1923, along with that of 26 November, deals in detail with questions and problems of apiculture. The loss of instinctive wisdom should be remedied by exact knowledge, thus the prompting to develop this thoroughly. This is the basis for finding new practical methods that incorporate underlying knowledge. Steiner offers ideas for remedying problems that have arisen.

As an example of the loss of an instinctive approach, he mentions the change from straw skep to wooden boxes, as well as the question of location. We can say of the skep that its curves are very well suited to a bee colony, but nowadays it is only occasionally used for demonstration purposes. The mobility of moveable frames has advantages that no one now wishes to relinquish. Rectangular straw hives are also sometimes used, though these are more laborious to make and maintain. It is known, however, that straw provides a different internal climate; the straw walls are more permeable and have better water-storage properties. Rye straw, in addition, has a high silica content, which in turn has an affinity with the 'hexagonal forces' of which Steiner speaks.

The question of where to locate hives is still one of the biggest debates in beekeeping. Manuals are full of good tips about how to choose the right location. In fact, the quality of any particular location can only be determined by setting up your hives and seeing what happens. Prior knowledge has proven of little value here. Hans Wille (1922–2002),[48] who ran the Centre for Bee Research in Bern, Liebefeld from 1957 to 1987, studied why bee colonies develop in different ways. According to his findings, the type of hive, the species of bee, the beekeeper and his diverse methods, have no effect on this. But very great divergences arise due to location alone. Ultimately the quality of a location is hard to explain, yet it is decisive for the success of an apiary, and is a key factor in the health or sickness of bees.

The new, modern hives and the question of location leads Rudolf Steiner to discuss the health of bees. He considers that the right relationship between 'blood' and 'gastric juices' is decisive for smooth conversion of nectar into honey. It is not completely clear what he means by gastric juice, but this relates primarily to a healthy composition of the bee's blood.

The idea of 'compensation' is important in this context: seeking ways of remedying deficiencies, which in turn requires creative solutions. As an example Steiner cites the idea of cultivating plants near the hive, for instance in a greenhouse. By growing certain plants that otherwise do not blossom at a particular season, one can make pollen available to the bees for purposes of sustenance and self-healing. But here he adds the basic requirement to engage with such questions in a natural yet forward-looking way, rather than trying to put the clock back. Purely intellectual, analytic reason brings us only fragmented insights that fail to grasp a whole context. To approach the nature of the bees, or to sense the qualities of a possible hive location, we need to be able to imagine ourselves into the bees' reality through holistic feeling and perception. Recalling Goethe's prompting to develop diverse methodologies, one can, as in an artistic process, use immediate, experiential understanding and judgement when seeking solutions to problems. This may allow us to redress the loss of old, instinctive knowledge and traditional approaches.

This idea of compensation is followed by the question of American clover, which illustrates the limits of innovation. Is it sensible to compensate for a lack of indigenous flowering plants by introducing non-indigenous ones? Steiner warns against cultivating plants from other countries or importing foreign bees. Today, of course, this has happened on a large scale, and we scarcely have the option to defend ourselves against such an influx—plants from all countries have long since been introduced here. Not everyone knows that the best-loved bee pasture crop, phacelia, originally came from abroad. Many problems with invasive neophytes, plants such as solidago or impatiens, are well known however. They injure the whole fabric of nature by suppressing indigenous plants, and the creatures connected with them, thus damaging species diversity. Absurdly, such plants serve bees as a partial compensation for the loss of indigenous flowering plants. Citing the example of American clover, Steiner points out that all this can be a problem for bees.

But indigenous bee species themselves have been widely suppressed too, through the uncontrolled import of bees from the mid-twentieth century onwards. Today great efforts are being made to regenerate and sustain our original black bee.

Since Steiner's day, our environment has come under much more stress, with corresponding efforts to regenerate it and at least keep some of the harm in check. In relation to bees this means trying to strengthen their natural capacities and resilience.

It strikes me as a really wonderful thing that the bee 'sucks' from the natural world this very valuable and life-enhancing substance of nectar and converts it into honey. Now you will understand what this whole honey-creating process involves if I describe the same process, though in a completely different form, amongst neighbours or relatives of the bee: in wasps. The wasps do not produce honey that can be used in the same way by humankind, although what they do produce can be very usefully employed in medicine. Yet the way the wasp works is very different from the bee. Later, or next time, I will talk about ants too. But first let us consider a particular type of wasp.

There are wasps that lay their eggs on plants or trees, for instance in the leaves or bark of trees.[49] Some even lay them inside tree blossoms. An egg deposited on a leaf looks roughly like this:

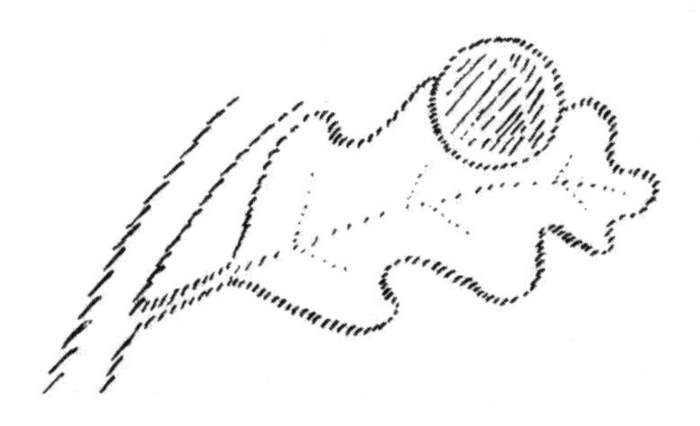

Here is the branch, and here an oak leaf, say. And now the wasp lays her egg into the leaf with her ovipositor—a kind of hollow barb. Now what happens? The whole leaf tissue around the wasp egg alters. The leaf would have grown quite differently without the egg. The growth of the leaf changes completely around the wasp egg and gives rise to what is called an oak apple gall. You find these brownish oak apples on trees, caused by a wasp laying its egg there, so that the substance of the plant is altered around it. It wraps itself round the egg. The wasp egg would not survive otherwise, if it were laid at random somewhere. It only thrives by virtue of the protective substance that grows around it, stolen in a sense from the plant by the gall wasp. The wasp takes this from the plant.

The bee lays its egg in the comb, and it develops into a larva and so forth, then hatches as a bee and later 'steals' plant substance, assimilating and converting it within itself. The wasp does this somewhat sooner. As soon as she lays her egg the gall wasp extracts the plant substance she needs. Thus the bee delays doing this compared with the wasp. The wasp does it sooner. In higher animals and the human being, the egg surrounds itself with a protective envelope already in the mother's womb. What is here extracted from the mother corresponds to the plant substance taken by the wasp. This oak apple is simply developed from the plant in the same way that the chorion forms in the womb to surround the germinal egg.[50] Later it is expelled with the afterbirth.

You see how the wasp works closely with the plant. In regions where there are many wasps, you can find these galls everywhere. The wasp here coexists symbiotically with the trees, dependent on them. Her egg could not survive or thrive at all if it did not form this protective envelope around it from the substance of trees or plants. Such a thing can appear differently too. There are galls that do not resemble apples but look like hairy excrescences, interwoven like this [see drawing].[51] But again, the wasp germ is at the centre. Sometimes you can find these galls in the form of small, shaggy nuts. This all shows how the wasps coexist with plants. Once the wasp has hatched, she uses her mandibles to bore through the gall and emerges as a wasp. In turn, after living as a mature wasp for a while, she will lay an egg in another leaf or similar. Thus egg-laying by these wasps is always in association with plants.

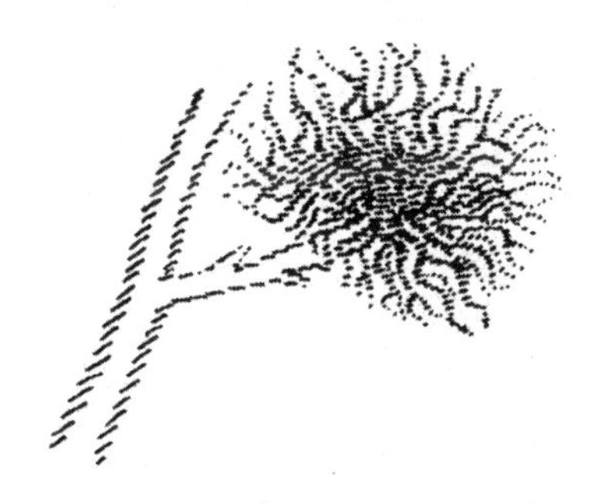

You might ask what this has to do with making honey. Actually it has a great deal to do with it; and we can learn how honey originates if we consider this. In older folk wisdom there were instincts that recognized what is at work here. You may know that in southern regions, especially in Greece, fig cultivation was an important pursuit.[52] Now there are wild figs that have a touch of sweetness but not enough to satisfy a sweet tooth. And the following was the custom. Picture a wild fig tree: a certain type of wasp is very attracted to it and lays its eggs in it. So imagine a wild fig tree, and a fig hanging from it into which the wasp lays its eggs.

The fig breeder is a cunning fellow. He cultivates these

wild fig trees on purpose, and lets the wasps lay their eggs in the figs. When the wasp larvae are developing and before they mature, thus long before they hatch out but still during their larval stage, he picks two such figs. He takes a reed stem and uses it to tie together these two figs containing the developing larvae. Now he takes them to the fig tree he wants to improve and hangs the two figs tied with a reed upon it, with the wasps developing in them. What happens now?

The wasps sense this since the figs that have been plucked now start drying out, no longer being fed by the sap of their tree. The immature wasp larvae sense this. Even the egg will sense it. And this means that the wasps greatly accelerate their development. The breeder starts doing this in spring; he first lets the wasp lay her eggs. In May he picks these two figs and does what I described. Good heavens, thinks the creature inside the fig, I'd better hurry: the fig is beginning to dry up. The wasp hurries to mature, and hatches far sooner than it otherwise would have done. If the fig had been left growing on the tree, the wasp would have hatched in late summer, instead of which it hatches in early summer. In consequence the wasp has to produce a second brood; it lays eggs in summer still, whereas it would otherwise only have done this next spring. These eggs [of the second brood of the year] are laid inside the figs of the tree the breeder wants to improve, but these are 'late eggs' that never mature or hatch. They develop only to a certain stage. And what occurs in consequence? The figs in which the second brood is laid become twice as sweet as the other wild figs! This kind of 'grafting' is an improvement that makes the figs twice as sweet.

But what has happened here? Wasps are related to bees, though different of course. Already at egg stage they extract from the plant what can become honey. The cunning fig cultivator picks two wild figs containing wasp eggs, binds them together with a reed and hangs them on another tree, persuading the wasps by this means to incorporate into their new host tree what they have assimilated from the original tree; and thus he gets them to introduce added sweetness into this new tree. Through the wasp, finely distributed sweetness is instilled into these improved figs; this occurs by a natural route.

Nothing has been taken away from nature here: the quality of honey has been augmented in the plant. A wasp cannot actually make honey like the bee does—it doesn't have the appropriate organism to do so. But when we compel her to lay a second brood in a year, which never matures, she transfers sweetness assimilated from the first tree to the other, improved fig tree. It then contains a kind of honey substance. This is something very distinctive, as you can see. These wasps have a body unable as such to extract nectar from nature and convert it into honey within them; but within nature itself they can enable a kind of honey to be passed on from one fig tree to another.

So the bee is a creature that develops from a wasplike body a capacity enabling her to accomplish separately from tree and plant what, in the case of the wasp, must be left within the tree itself. We must say that the bee is a creature that retains within her more of the power the wasp has only while an egg or larva. In maturing, the wasp loses this honey-creating capacity. The bee retains it and can exercise this power in maturity. Now consider what this means: to look closely into nature and say that the sweetness, the nectar in plants, can be augmented by supporting a natural process; by transferring at the right time the wasp to the tree one wishes to improve.

Here in our own regions the following can't be done. It can't be done any more at all nowadays. But there was a stage of earth's evolution when it was possible to gradually breed bees from wasps by drawing on this technique, used today by clever fig breeders, and two thousand years ago already, to compel wasps to lay a second brood in figs. You see, the bee is a creature that was bred from wasps in very ancient times. But today we can still see how, through this life of the wasps, sweetness manifests in nature itself.

And consequently you can also see what is at work when bees deposit their honey in the comb. The comb of course consists chiefly of wax. In fact this wax is not only needed for depositing and storing honey but the bee can only produce honey when her whole body works in the right way. Thus she has to secrete wax.

Well, the second fig tree, in which sweetness is naturally increased, is also richer in wax than the wild fig tree. In fact this wax enrichment is the very thing that distinguishes the improved fig tree from the wild one. Nature itself enhances the wax content. Thus the improved fig, the sweeter fig, grows on a tree that inwardly enriches its own wax content. This is a kind of prefiguring of what occurs in apiculture.

And if you set to work very precisely, you can take a cross-section of the stem of an improved fig tree and, examining it, will find forms resembling wax cells [see drawing]. Forms resembling bee comb develop from this wax deposited in the stem. The improved fig tree becomes richer in wax, and within the stem the wax arranges itself into a kind of cell form. And so, if we reflect on this manner of improving figs, we have a honey culture that remains within nature rather than being extracted from it.

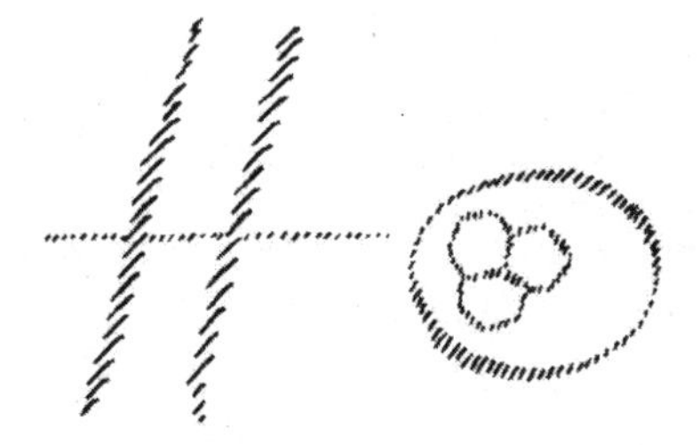

If I can put it like this, the bee draws out into the public domain what remains enclosed within nature itself in the fig. The bee draws this into the open. And as a result [of the fig improving] the wax that otherwise remains in the stem, and there forms a type of natural cells—not so marked, not so compact, only intimated—is driven up into the fig. The whole wax- and honey-forming process is then contained in the fig. Thus a kind of apiculture occurs, instead, inside the tree itself, so that all nature is a beekeeper.

What does the bee do first? She lays an egg in the comb, within the hive. The egg develops and matures. There is no need here for the bee to transform plant substance into a gall but she takes nectar straight from the plant. Nor does she then fly to another tree to enrich its wax content but, by itself, makes the cells, the comb, that otherwise form in a stem or trunk, and into this deposits the nectar converted into honey. In the improved fig, by contrast, the sweetened sap pervades the whole fig tree. So we can say that here something is accomplished openly and distinctly that otherwise in nature remains within the tree and is accounted for by the interplay of wasps and trees. And this makes very clear what you actually have before you when you look at a hive with its artistically structured comb of wax cells. This is really a wonderful sight, isn't it Mr Müller? What a sight: this artful creation of comb cells. And these cells with honey inside them!

Yes, take a look at this. And then suddenly it will strike you: good heavens, the wonderful wax cells of the bee comb are a sort of artfully constructed tree trunk with its branchings. The bee does not introduce its eggs into a tree but creates something like an outward picture of the tree's structure; and instead of having the fig grow from it, the bee deposits honey into these prefabricated cells. Thus the bee creates a kind of image of the artfully improved fig tree.

This insight into the inner workings of nature can teach us how to learn from nature herself. But we must be able and willing to do so. Human beings will learn a great deal still from nature, but to do so they must first perceive the spirit in nature. Otherwise we'll just stand there gaping, won't we, when we travel to Greece and see a cunning fellow binding two figs together with a reed stalk and then casting this up into a wild fig tree. Even if we're scientists, we'll stand there gaping, at a loss to understand what is happening. In fact he is saving nature its beework, getting it to put honey straight into figs instead. And in the regions where they thrive, figs have similar health-giving properties to honey, since a kind of early stage of honey is present in them.

These things need to be considered if we're discussing something as important and far-reaching as apiculture. It seems to me that such reflections will gradually enable us to gain more accurate insights.[53]

R. Hahn: After the lecture, I asked Dr Steiner about the cause of foul brood. He said that he could only say something specific about this disease once he had investigated it properly. But he thought that foul brood was probably to do with a uric acid disorder in the queen bee. Then he said, 'You see, the bee too has uric acid in her organism, and the cause of this problem will be an imbalance in its composition.'

Regarding the breeding of the bee from the wasp, mentioned in the same lecture, he said roughly: 'This process occurred in ancient Atlantis, when the diverse animal forms were not as fixed and finished as today, and there was not yet such a clear demarcation between different species. Today this kind of breeding would no longer be possible.'

Origins of the bee

Are bees domestic creatures or wild ones? That must remain an open question at present, since no difference exists between a domestic and wild species. In itself this suggests they are wild. Yet human beings dwell in close proximity with bees, which inhabit man-made hives, often close to human habitation, and connected to us through intensive give and take. We have already mentioned the possibility of a close soul connection. And there is much to suggest, also, that the bee is a domesticated creature or at least a part of our agriculture. As far as the origins of honey bees are concerned, we assume that they evolved alongside flowering plants, which appeared around 100 million years ago. There are bees preserved in amber dating back 60 million years, whose form is identical to that of our own honey bees. These could be solitary bees, since the discoveries tell us nothing about possible bee communities at that date. We know nothing at all about the origins of bee colonies.

And yet Rudolf Steiner's account of the breeding of bees from fig wasps makes many shake their heads in disbelief. This is a stumbling block even if Steiner replies to Mr Hahn's further questions on the subject by saying this took place at the time of 'ancient Atlantis', commenting also that species were less clearly delimited and differentiated in those days. It cannot at least be entirely excluded that bees were bred over time by human beings from a primordial form of insect close to wasps.

At this place in the lecture it is also noteworthy that sugar and nectar formation in plants is connected with wax formation and the bees' comb-building activity, as well as silica forces and their tendency to form hexagonal crystals.

In other contexts, too, Steiner speaks of the origins of the honey bee in early times.

Now let us return to the first Saturn condition, when the human being was a being of spirit and soul, whose body was in general always the same—a being that knew himself to be immortal at a low evolutionary stage, and continually transformed his body. We see this condition still in a creature with a remarkable communal life, whose group soul is in some respects at a higher level than the human being. I'm thinking of the bee. We must think of the whole colony differently from the single bee. [...]

It is very interesting to trace a parallelism that physical science can tell us little about. What do we still retain in us of the Saturn condition? Warmth! The warmth of our blood. The warmth which was distributed everywhere at the time of Saturn separated out and today forms the warm blood of humans and animals. If you measure the temperature of a bee hive, you find it to be roughly the same temperature as human blood. So the whole hive develops a temperature that corresponds to our own, and it can be traced back to the same evolutionary stage as human blood. The occultist therefore says that the bee is born from warmth, is a warmth creature, in the same way that he regards the butterfly as born from air, an air creature, the fish as a water creature and the snake as an earth creature.[54]

> Warmth within the hive, as already described, reveals an evolution not found in any other insects. It is hard to understand how this comes about, particularly since there are no transitional or parallel phenomena in the insect world. One can gain a sense that this evolution originates from elsewhere. In Steiner's text we hear of the bees' relationship to human beings in ancient conditions of the earth, but their origins remain a mystery—many aspects of the life of bees remain hard to classify.

7. BEES, WASPS AND ANTS

What is the relationship between bees and flowers? What connects them? And what is the significance of honey for human beings? Then there's also a further question about the laying of bees' eggs [oviposition].

Right, let's speak about these things again today. The thing is this. We have this fertilization of the queen during the nuptial flight. So first the queen has been fertilized. Then we need to consider the period from the depositing of her eggs to the emergence of the fully fledged bee. This period takes 16 days for the queen, 21 days for the worker, and 22 to 24 days for the drone. So these three types of bee are different from each other in this respect—that they take different periods to hatch. What underlies this? A queen develops from a larva especially by virtue of the fact that the bees feed the larva in a particular way. Queens are fed a different substance, and this hastens their growth.

Now the bee is a sun creature, and the sun takes roughly the same time to complete one revolution around itself as a drone needs to develop.[55] In other words, the queen does not wait for a full revolution to be completed before maturing; she stays within the phase of a single revolution, and in consequence she remains entirely subject to the sun's influence. As a result, she becomes a bee capable of laying eggs. Ovulation is a capacity that is also subject to the sun or the cosmos.

The moment the larvae are fed in such a way that they mature by the time the sun has almost completed a whole revolution, then they fall more under the sway of earth development. As the sun concludes its revolution, the bee begins to fall more under the sway of earth influences. The worker bee is still very much a sun creature, but just beginning to have earthly qualities. And the drone, which takes more or less the period of a complete revolution to develop, becomes entirely earthly, and thus separates itself from the sun.

So we have three things here: the sun queen; the worker bee, still with extra-terrestrial forces; and the drones, which no longer have sun qualities and have become earth creatures entirely. Everything else that occurs is not subject to the sway of earth forces, only fertilization itself.

Now the remarkable thing is this. Consider the nuptial flight. Lower animals don't take any particular pleasure in fertilization, and may even avoid it. There are many examples of this. Therefore the nuptial flight is really a flight on which the queen embarks towards the sun, for it doesn't occur when the sky is overcast. And the drones, seeking to introduce terrestrial qualities into solar ones, even have to battle in the air. The weaker ones fall back. Only the strongest, who can fly as high as the queen, can accomplish fertilization.

But simply because the queen has been fertilized, this does not mean that every egg in the queen is fertilized. Only a portion of them are, and these become the workers or queens. A portion of the eggs are not fertilized, and these become drones. So if the queen is not fertilized at all, only drones emerge from her eggs. After the queen is fertilized, drones emerge [from her unfertilized eggs], or otherwise worker bees and queens if the egg is fertilized, and thus the heavenly realm comes into contact with the earthly.

Here again the drones are most exposed to earthly forces,

because the eggs they come from are not fertilized. To remain viable they need a full exposure to earthly forces; they have to be fed for longer and so on. Does this answer your question?

Some years ago I heard that bee or wasp stings can help with rheumatism.

Here I return to a question which may not have been answered last Monday. Mr Müller told us about a man who had some kind of heart condition, it seems, and who collapsed after being stung by a bee,

Mr Müller: The doctor advised him to give up beekeeping, as it might kill him!

The heart condition testifies to the fact that this person's I organization was not properly integrated, and here we must recall what you know from other lectures I have given. As you know, we distinguished four aspects of the human being: firstly the ordinary physical body, which can be seen and touched; secondly the etheric body; thirdly the astral body; and fourthly the I organization. This I organization acts upon the blood, in fact impels its motion; and as the blood is driven in this way, the heart beats. Books on medicine and physiology give a completely false account of what happens: they describe the heart as a pump, which pumps blood everywhere through the body. That is nonsense; in reality the blood is not pumped but impelled by the I organization itself, and thus brought everywhere into motion.

If someone claims that the heart drives our blood it is like saying that a water turbine drives water. Everyone knows that the water drives the turbine. Similarly in us, the blood is impelled and drives the heart. Except that here the blood first flows in and, as oxygen unites with carbon dioxide, it flows back again, thus swirling first forward then backward, giving rise to the heartbeat. It is therefore true to say that the I organization intervenes directly in blood circulation.

Now this I organization is actually mysteriously contained in bee poison. The force circulating in your blood also exists in the bee sting. And it is interesting that the bee needs the bee poison inside herself. She needs it not only so that she can sting—her stinging ability is merely a chance addition. The bee needs the bee poison inside herself because she requires the same power of circulation that we human beings possess in our blood.

As I said, the hive is like a whole person. Now imagine that you get bee poison in your body, that is, in your blood. Like any poison entering the body, it passes into the blood. As a normal person, this will set your blood moving faster, giving rise to inflammation; but your heart will cope with it. But if someone has a heart disorder, the strengthening of his I organization by the poison impacts on a somewhat weakened heart valve, and he can faint or even die as a result. That is the case recounted by Mr Müller.

Now the remarkable thing is that everything that can make someone sick or kill them can equally cure them. And this is the great responsibility we have when preparing medicines: there is no real remedy that will not, when mistakenly applied, produce the same illnesses that can be healed by it. So if bee poison can induce fainting or even death, what is happening here? You see, if someone faints, the astral body and—especially—the I withdraw from his physical body, depart from it as they do in sleep. In sleep this is healthy, but when we faint it is pathological. In fainting or sudden loss of consciousness, the I remains stuck, whereas in sleep it withdraws fully. If someone has a weak I organization, he cannot bring the I back inside him. You have to shake him and shake him so that he wakes up, so his breathing recovers and so on. You have to employ artificial means to revive him. In such cases, as you know, you have to cross the person's lower arms over his chest, then uncross them and cross them again: this is to revive the breathing if someone faints. This artificial respiration always involves the attempt to bring the I organization properly back into the organism.

Now if someone has rheumatism or even gout, or other deposits in the body, we have to try to strengthen the I organization. For why does someone get gout or rheumatism? Because the I organization is too weak and does not bring the blood into sufficient motion. The blood needs to be stimulated. If the blood is not in proper motion, for instance is flowing too slowly for the organism in question, small crystals are deposited everywhere, and pass into the vicinity of the blood vessels. These small crystals are made of uric acid and fill the whole body, causing gout or rheumatism, because the I organization is too weak.

If I now give this person the right dose of bee or wasp poison, this strengthens his I organization. But I must not give too much, for otherwise his I organization cannot assert itself. But if you give just the right amount to strengthen the I organization, bee or wasp poison can be a very good remedy. It must however be combined with other substances. Such things are done. For example, there was an old, traditional remedy called 'Tartarus',[56] though made of other substances.

These toxins can always be used in medicines, as here for instance to strengthen the I organization. But before prescribing it you need to know the particular patient. Let's say someone has gout or rheumatism. The first question is this: Is his heart sound; does it, in other words, function well under the sway of blood circulation? If this is the case, you can use bee or wasp poison to cure him. But if his heart is unsound, you have to enquire further. In the case of nervous heart disorders, this will not matter too much; but if someone is suffering from a heart disorder resulting from a valve deficiency, you'd have to be very careful about giving such a remedy. It would impact strongly on the valve, on the heart. If there is a valve disorder, it is possible you could not employ bee or wasp poison at all. That is why it is so dangerous to say in general that some medicine or other is good to combat this or that disease. Certainly one can make a remedy to which is added wasp or bee poison—we actually have such a medicine[57]—and add binding agents, gelatinous or herbal binding agents, and put this remedy in an ampoule for injection. It is injected, imitating the bee's sting; but an actual bee sting elicits a far greater reaction. And one can produce this remedy, and can call it a medicine against rheumatism.

But above and beyond this we must be concerned as to whether the patient will tolerate the remedy, and this in turn depends on the general state of health of his organism. The remedies that work more deeply into the organism should really only be given after checking the patient's overall state of health. If you find medicines that are universally praised for a particular condition, they are usually ones that can do little harm, and nevertheless bring benefits. Such medicines can be sold over the counter. We can even accept the fact that unpleasant consequences sometimes ensue, since a cure always involves unpleasant consequences. A patient will always need to slowly convalesce and recover from the illness.

Today many people seek a cure although they are not ill. Before the war this was even truer than nowadays, when there is such financial hardship in some countries. Doctors are having a hard time there since people are not seeking cures as much as they did. Before, everyone wanted to be cured—whether healthy, only mildly ill, or really ill, or gravely ill. Nowadays only patients with very serious conditions can afford to seek a cure!

If a very strong fellow gets rheumatism—it is usually no such thing, but rather gout-type conditions—a bee sting, as Mr Burle said, can be of great benefit. He can be cured because he tolerates the reaction well.

And usually an ordinary person who gets rheumatism can of course cope fine with the right dose of bee sting given him as a remedy, and this can cure him. But an actual whole bee sting will usually cause so strong an inflammation that this must first be treated, and the bee poison if possible removed quickly; and then not much of the bee poison will be left over to cure the rheumatism. This will be the case at any rate in normal subjects.

But let's take the following case. Rheumatism can occur as follows: a person doesn't work much but eats a great deal. He will usually have a very healthy heart if he works little, eats a lot—until things start to be a cause for concern.

The heart is a highly resilient organ, and is only very gradually spoiled from within after many years, unless there were already inherited problems or it was overtaxed from an early age. But someone who eats a great deal may also drink a lot of alcohol with his meals. In consequence the I organization is stimulated, and blood circulation becomes very intense. In this case the heart and its pulse can no longer keep pace. Toxins, uric acid and so on are deposited everywhere.

The person's heart may remain perfectly strong for a long time, but he will have gout and rheumatism everywhere in his body. In some circumstances a bee sting can be extremely beneficial for a person like that.

Mr Burle: I'm not sure if the person this was said of was at the same time also a bit of an alcoholic.

You mean you didn't enquire? Especially if you're using a remedy like bee sting—which is a strong remedy—you have to take very great care to consider the patient's whole state of health.

Mr Müller says that he got rheumatism after a chill; he used sunshine as a cure and it disappeared, but it has come back a bit this summer. He also believed that the condition could be treated with bee stings; but on one occasion he had an unlucky day when he was stung on both legs—he had about 32 bee stings. The only problem resulting from this was that he went all the colours of the rainbow for a week. Nor does he always swell up when stung. The human body differs from person to person: one person can die of a bee sting whereas he, for instance, has had up to 60 stings without his heart beating any faster. One person is just more resistant than another.

When you were stung so many times, had you already been working with bees for a long time?

Many years!

You may have forgotten the first time you were stung. The first time, someone will more or less feel the full effect. The man you were telling us about was surely stung for the first time. And once you have this poison in your body, in your blood, you become increasingly able to counter it, increasingly immunized as they say. So if someone was stung a little when he began working with bees and otherwise has a healthy heart, he will become ever less susceptible to bee stings. If you know you're healthy you can even make sure you're stung by a bee so that further stings are less severe. You'll get the symptoms—rainbow-coloured skin and so on, but this is more external. The blood is immunized. This depends not only on your organization but on what has previously been introduced into your blood. I'm surprised that the doctor who observed this case did not tell the man that if he were stung a second time it wouldn't be so bad; and by the third time he would be immunized. But perhaps he had a severe heart disorder, and this would have been too dangerous. This also has to be considered.

There are doctors today who think every beekeeper should be vaccinated before he starts working with bees, but this is a dangerous path to go down. When people go to war they are filled with all kinds of poisons, and this cannot be recommended. The blood can become a kind of refuse dump; it deteriorates from all these substances introduced into it. This rebalances itself after a while; the blood becomes healthier again, and is protected against new toxins of the same kind.

Mr Müller: As far as drones are concerned, and the three types of egg, Dr Steiner has covered everything except perhaps for one point that he may not be aware of. When the bee colony seems to be perfectly healthy, times arise when the queen is of lesser quality, or when she grows too old, when all the eggs she lays become drones. His 30 years of experience have taught him that a queen of poor quality, either because of illness or age, is still able to lay one or two good eggs, but that the majority will be drone eggs.

Then he has a question about the bees' honey production, how they produce it and whether the beekeeper needs to help by giving sugar. The comments made here suggest that the beekeeper should not use sugar at all. If you feed sugar during the honey-producing period, you get a black mark—like a worker who makes himself unpopular, and the employer wants to get rid of him. He himself has had bad experiences with honey brought in from abroad.

It is quite right to say that it won't be the same product if you use sugar artificially. And if someone wishes to eat sugar as well as honey, he can do so of course. You should not put water into wine because you think strong wine is bad for people, since they should get what the label says on the bottle. And here likewise the best thing is for beekeepers to organize mutual checks themselves since they are the ones who know most about it.

As concerns the drones, I'd also like to say this. You may suspect that the queen was not properly fertilized and that too many drones are hatching. Then, if you don't want to leave things to the bees themselves—and if too many drones are hatching, as has been found in trials, the bees won't correct this—you can use especially strong feed to make sure they hatch earlier, not at 23 or 24 days but at 20 or 21. Then the drones can become more like worker bees, pretty poor ones, but similar to them at least. That won't be a long-term solution, but it shows you what part is played here by the larval and pupation period.

These are things of course that are probably not done in practical beekeeping, but they are theoretically possible. What is fed to the bees has a very strong effect; and it cannot be disputed that in some instances it has been possible to make a worker bee into an egg-laying one, although she isn't a proper queen.

All this shows how flexible is the nature of such a creature; but it won't have much effect on practical apiculture.

Mr Müller: We call that a virgin queen—it's a disease in the hive.

It is not of much importance in practical beekeeping. But in the hive the colony tends to be able to use particular feeding methods to make, say, an egg-laying bee from a worker. Yes, that's a kind of disorder. The hive is a unity, and so the hive is sick. It is just like force-feeding a goose, something that particularly develops its liver forces: the liver becomes over-healthy while the whole organism becomes sick. If you induce a worker bee to act as a queen, she is, really, an over-healthy worker bee, but the hive must then be regarded as sick.

Further questions may occur to you at some point. We can always come back to them. But now I'd like to say a few words in relation to Mr Dollinger's question.

We can make precise distinctions between insects that are bee-like in the broader sense: bees, wasps and ants. They are related to each other; and last time I spoke about the interesting matter of gall wasps that lay their eggs in trees and so forth; and I showed that these wasps perform a kind of inner manufacturing of honey. But other types of wasps are similar to bees in a different way, in that they build a kind of comb.

For instance, there is an interesting species of wasp that builds its nest as follows. They fetch tiny bits of tree bark and suchlike, which they bite off, then mix this with their spittle and first make a couple of posts from the substance, attached, say, to a stiff or rigid leaf. They keep repeating this, mixing the substance with spittle and making from these posts something that appears very similar to one cell of a bee comb. But if you examine the substance it contains no wax. The wasp cell is greyish, and the material is very similar to our own paper. It really is like paper pulp. The wasp attaches to this first cell a second, third and fourth, hanging on the leaf. Then it seals them up after eggs have been laid inside. And as the egg-laying is still going on, the wasp makes from the paper a curious kind of loop like this [see drawing] and then a kind of lid: on one side the cell is left open, as an entrance to fly in and out of, so as to attend to the larvae.

Then it adds further cells, repeats the same thing, attaches a loop and then again a lid, with an entrance here. The structure can become very extended, like a pine cone. It builds this comblike structure which resembles a pine cone, consisting only of the paper pulp.

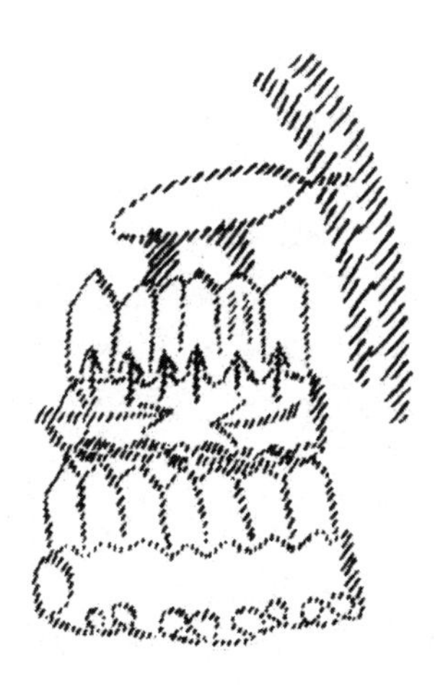

As you know, other types of wasps build nests enclosed with a kind of outer casing. Wasps' nests can have all kinds of shapes.

But now let's consider what is really happening here. If you ask me what the bee does in order to make its comb from wax, I have to say this. The bee flies to flowers or to blossoming trees, and is less concerned with woody parts of the plant or bark. Basically she seeks out flowering parts, at most the parts of a flower that tend towards leaf, but that a good deal less.

Now there are instances where higher insects like the bees seek out something other than flowers; not woody parts either, but something different that can taste very good to them. Not so much bees, but the wasps do it, especially, and ants even more so. Whereas they use harder plant constituents, woody parts, for their structures, wasps and ants, in contrast to bees, are especially interested in the juice that comes from aphids. This is very interesting. You see, the harder the substance they use for their nests, the more they delight not merely in nectar but in a small creature sitting on the flower, and closely resembling it, the aphid. These are very noble creatures—forgive me if I now resort to ant-speech. I would not use the term in human speech. To the ants an aphid is a noble creature, is entirely flower-like. The juice it secretes is really [for the ants] the finest honey you can imagine. Wasps really consider aphids rather a delicacy.

But when we come to the ants, we find they do not have the strength to create a nest like that of the wasps. The ants do it differently again, piling up layers of earth, within which you find passages everywhere, a whole labyrinth of them. And through these passages the ants drag all that they can use of the harder constituents of plants, bits of bark and so forth. But ants especially love what is already dead matter in the wood. The ants seek out what they need in order to extend this structure made of layers of earth, using wood from a tree that, say, has been sawn down, leaving the stump. They fetch the core and the bark and use it to extend and elaborate their nest.

Bee: flower sap: wax cells
Wasp: flower sap, insect juice: harder cells
Ant: insect juice: no cells

The ants therefore use the hardest substance of these three for their nest, and do not manage to create cells. The material they use is too hard for this. Bees use the material contained in flowers themselves, make their wax cells from this, and for their food rely on the nectar they obtain from flowers.

The wasps use a harder material, creating a structure out of something like paper: a harder material than wax, but thin and therefore more brittle than bee comb, but harder in itself.

The wasp treats aphids as a delicacy, but still also feeds on what plants contain, like bees. But the ants, who use such hard material in their nest so that they can only build passages and caves, not combs with cells, delight especially in aphids. Ants sometimes take whole aphids into their nest, where you can find them if you open it up.

This is very interesting. If you go to a village, behind a row of houses you'll find sheds with dairy cattle inside, which give people milk. The ants do something very similar.

Everywhere in ant nests you find little structures containing aphids—the ants' cows, though at a correspondingly lower level. Inside these little 'sheds' are aphids, and the ants approach them, stroke them with their feelers, thus delighting the aphids which then secrete their honeydew, which the ants can imbibe. They gain the most important part of their diet from this aphid honeydew, obtained by 'milking' these creatures. Cows are similar in that they need to be stroked too. The ants herd aphids on the flowers and leaves where they settle, and take very good care of them.

So we can say it is an excellent thing that aphids exist—a fine thing indeed that aphids are found near ant heaps. The ants carefully herd and collect the aphids and use them further in their little barns. How wonderfully inventive nature is we see from this aphid cattle-herding in the insect world.

Now the ant, which uses such a hard material in its nest, cannot make do really with mere nectar or plant juice. For its food it needs instead what another creature has already made of this juice. The sap has to have passed through the aphid. And so we can say this: pure nectar in the case of the bee; in the case of the wasp, plant saps and insect juice, and harder cells; and in the case of the ant, insect juice only as food, and therefore no comb cells. The ant no longer has the strength to build cells, and besides what it obtains from flowers it must also have this additional food from its little cow-sheds, otherwise it cannot survive.

So you see what interesting relationships exist between flowers and the insect world. The bees need pure nectar. The others, the wasps, and the ants particularly, rely for their food on the passage of this nectar through the aphids. In consequence they can use materials for building their nest that are no longer directly drawn from flowers.

There is a very great difference between the wax comb of bees, the paper nest of wasps, and the more external edifice of the ant nest, without cells. The great difference here lies in the food of each insect.[58]

In this lecture the focus is on social insects, on the one hand their poison and its effects on humankind, and on the other the different nature of bees, wasps and ants.

Bee venom

The effect of bee venom is something we can all observe in ourselves as long as there is no risk of an allergic reaction. The latter appears within a few minutes, accompanied by dizziness, nausea, blotchy rash and sometimes difficulty breathing. If such a reaction occurs, seek medical help immediately. Otherwise, to test the effects of a bee sting on you, wait for a chance sting or grab a bee and place his abdomen and stinger on your skin. The notes below follow from a test I did on myself, trying to observe both my immediate experience and any changes in bodily sensations and inner feelings: 'I was struck by a sense of wakefulness: things grow brighter, one becomes more awake, and there's a burning sensation. Physically, you feel warmth, heartbeat intensifies. The pain makes you feel uncertain, but as it wanes there is a sense of being well-centred, and of physical verticality, along with an astonishing awareness that reaches right to the tip of your toes—attentiveness to the whole body. Inwardly this dynamic expresses itself in attention to one's own I, and one's surroundings. A sense of preserving and cultivating what belongs to me, and attending consciously to my boundaries, as well as experiencing those of the other.'

This self-observation is different in some respects from the homoeopathic picture for Apis, but has some affinities with Rudolf Steiner's indications. Firstly there is the striking effect on circulation, heart and blood. Through this experience we can easily imagine that bee poison, via the warm, moving blood, is able to prevent the accumulation of uric acid crystals. In certain regions of the body, blood warmth can be markedly reduced, making it more possible for such deposits to settle. The action of warmth can be healing here. In several places, Rudolf Steiner mentions bee poison in connection with the unity of the hive. Feeling one's own inwardness, and body-soul boundaries, is similar to the process he describes. The experience is also reminiscent of how the I organization deposits silicic acid at the body's peripheries. One fascinating passage of this lecture, in particular, can be studied more closely in this respect: it intimates a truth which is however beyond merely rational understanding.

Now the fact is that this I organization is really contained in a mysterious way in bee poison. The vigour you have in your circulating blood is also contained in the bee sting. And it is interesting that the bee needs this poison within her. She doesn't need it only in order to sting, which is a secondary aspect. The bee has the poison in her because she needs the same power of circulation that we possess in our blood.[59]

The social insects

In relation to the social insects, Rudolf Steiner describes the connection between the food of the insect community and the way it builds combs or nests. Bees draw their food from blossoms: nectar supplies carbohydrates, and pollen gives the protein that bees need. Thus bee life is entirely dependent on flowers. The strength they draw from this food enables them to build their comb from wax. This capacity to build the comb from their own bodily substance is connected also with the production and regulation of warmth referred to in Chapter 6 (pp. 116f.). Alongside warmth, the wax they secrete themselves contributes to the independent status of the colony. It requires some kind of protective casing, whether a hollow tree trunk or hive, and bees prefer to be raised above the ground at a height of two metres. All these things—feeding from blossoms, the building of comb, and the preference for elevation—show bees to be creatures midway between heaven and earth, with only a little affinity to the ground and the soil.

The wasp has a broader and more earthly diet, giving its brood protein in the form of meat or carrion, and collecting rotting wood to build its papery comb-nest. She only rarely visits flowers. The years in which wasps reproduce abundantly are also rich in aphids, showing their dependence on these sap-sucking creatures. The wasp community dies out after the first frost, and a surviving queen founds a new colony in spring.

The ant, finally, lives in the earth, builds no comb and specializes in assimilating plant and animal substance. I have myself observed how the ants wait near their aphids and, on my approach, place themselves on the alert to defend their sap-milk herd. They can also gather them in some shelter, for instance in a fierce storm. But aphids only produce 'honeydew' when they are on a plant, broaching the veins of stem or leaf so that the sap, under pressure, flows into and through them. From this they filter out the proteins they need, letting the juice exude. This honeydew is something beloved of ants, wasps, but also bees. The latter, however, only harvest it in emergencies, when there is insufficient nectar. This is rarely the case, but when it is they gladly collect the drops the aphids let fall onto leaves or needles, and make from it the great delicacy of forest honey. As Rudolf Steiner tells us, they can also fall ill from this if compelled to feed in winter from this kind of honey alone.

8. INSECTS AND PLANTS

Today I will continue with the reflections that started last time from Mr Dollinger's question. If other issues come up we can deal with them too. To answer this question I started with a consideration of ants. Bees, wasps and ants are related species, albeit with very different kinds of behaviour. And we can learn a great deal from them about the balance of life in the world in general. The more we study these creatures the more we can realize what wisdom lives in their work and in what they accomplish.

Last time I told you how the ants build their nests, either as little hills made of earth or from small splinters of rotting wood, and other things they mix with them. Thus they make their anthills, their earth nests, inside which are numerous passageways through which they move in large droves, in processions. You can see them emerge from holes in the anthill and head off into their neighbourhood to gather whatever they wish to gather.

Sometimes, though, these creatures use something that is already there rather than building their own dwelling. Imagine someone has felled a tree, and leaves the stump in the ground. Along comes an ant colony and burrows into it, making chambers and passages. Perhaps they layer up earth too, make a passage and then another and so on. All the passages are interconnected, a whole labyrinth of them. The ants live there, passing in and out and fetching what they need from their surroundings for their nest and their food.

To say this is all simply instinct is well and good, but it tells us little. You see, if the creature finds no such tree stump, it creates a nest of sand or whatever it finds. But if it can save itself the work involved it will use something already there, like a stump. In other words, the insect orients itself to different conditions, so it is hard to say that it simply has a generalized instinct, which would mean that everything it did was generally dictated by this instinct. No, the important thing is that the insect accommodates itself to outer circumstances.

There are fewer anthills here, but as soon as we come to more southern regions, ants become really quite a plague. Imagine a house, and in one corner, unnoticed by the inhabitants, ants establish themselves and start carrying all kinds of things from the vicinity—grains of earth, small splinters of wood—and first build themselves a small chamber in some corner that has long been left uncleaned. And from there they make inroads into the kitchen, the larder, through all kinds of complex routes and passages, and fetch what they need for food and anything else. In these southern regions a house can become completely infested with them. Of course you may not even know that you're sharing your house with the ants until you notice that things are being eaten in the larder and discover the passages they're coming through.

Again, saying this is all instinct isn't saying much, since then you'd have to postulate an instinct to create a nest in this particular way in this very house. In fact it isn't only instinct; there is wisdom living in these insects.

But you won't discover that an individual ant is especially wise. The way it behaves when separated from its colony does not seem to contain any particular wisdom. The consequence is that instead of imputing insight to an individual ant you must ascribe it to the anthill as a whole. The whole beehive is similarly wise. The individual ants do not each have their individual intellect. The way they work together is extraordinarily interesting.

But even more interesting things can happen here. There is a type of ant[60] that behaves as follows. Somewhere on the ground it raises a kind of rampart or ridge [see drawing]. And then it forms a level circle of soil around the hill. The ants burrow into this hill. This rampart can also have a cone like a volcano, with passages inside it which extend into the surrounding area.

Now these ants do something very unusual. They clear from the surrounding vicinity all the grasses and plants except for one particular species—they bite them off and clear them away. Sometimes they clear everything so that you have a small hillock in the middle and all around it the ground looks as if it has been finely paved. This is because the soil grows more compact when they gnaw away all the plant cover. The soil becomes very compact. So here you have an anthill and around it a kind of paved area, very smooth like an asphalt road but lighter in colour.

Now these ants head off into their surroundings to bring back a particular species of grass, and begin to cultivate it. Whenever the wind brings other seeds, they gnaw off the seedlings the moment they appear, clear them away from the area they have made smooth, and allow only this one type of grass to grow there. So the ants establish a kind of estate, cultivating there the grass crop they have selected, and allowing nothing else to grow. This type of grass acquires a very different appearance than it does when growing elsewhere, in looser soil. There it looks very different. In this soil made compact by the ants, this species of grass they cultivate acquires seeds that are very hard, hard as gravel.

Yes, you can find anthills like this, surrounded by agricultural activity: these are agricultural ants! That's what Darwin called them.[61] He made a close study of them. So you find this farming activity all around, the compact soil and the seeds extremely tough. Then, when this has been done, the ants come and harvest their crop; they gnaw off the seed head, and bear it into their nest. Then they stay indoors for a while and you don't see them; but they're at work in there. They gnaw off everything they can't use, the chaff attached to these flint-hard seeds. And after a while the ants come out again and cast away everything they couldn't use, retaining inside the anthill only the flinty seeds. They use some of these as food, chewing on them with their tough jaws, and some they use to extend their nest. They are real farmers: they look closely at what is growing to see if they can use it or not; and what they can't use they dispose of. Human beings aren't very different in this respect. These agricultural ants are great experts in accomplishing what they need, and pay much attention to detail.

We can ask what is really happening here. In fact, they are cultivating a quite new species of grass! Gravel-hard rice grains as the ants cultivate them do not otherwise exist; they are only produced by ants, and then processed by them afterwards. What's happening here? Before we embark on this question, let's examine the matter from another perspective.

Let us return to the wasps,[62] creatures which as I said lay their eggs in tree leaves and tree bark, causing galls to form, from which in turn young wasps hatch.

But it can be different too. There are caterpillars that look roughly like this. You have all seen them—covered in thick hairs, prickly. The following can happen with a caterpillar like this. One or more wasps of a particular type come by and simply lay their eggs in this caterpillar. When these eggs are ready the grubs hatch from them—larvae that are the first stage of bees, wasps and other insects of this type.

The same is true of the ants. You know that if you clear away an anthill you find the small white 'ant eggs' as they are called. In fact they are not eggs but pupae. The eggs are small, and from these the larvae hatch. The pupae are wrongly called 'ant eggs'.

It is remarkable what happens when the wasp lays her eggs inside the caterpillar. I have described this to you before.[63] The grubs that hatch are hungry and there are lots of them in the caterpillar: they feed on the caterpillar's body; but something remarkable happens now. If the wasp larvae consumed the caterpillar's stomach, their larder would be exhausted and all the wasp grubs would perish. If any vital organ in the caterpillar were consumed—anything that serves it as eye, heart or digestive organ, for instance—the whole thing would come to an end. These tiny wasp larvae show the sense not to eat any part that the caterpillar needs to go on living; they only gnaw away at organs that can long be consumed without terminal injury. The caterpillar does not die, at most it is sick. But the wasp larvae can go on eating.

It is wise, therefore, that these wasp larvae do not touch a vital organ that could spell the end for the caterpillar. You may have seen what happens when the mature grubs emerge:

they crawl out, and the whole caterpillar was, really, the foster mother for them, she nursed the whole brood with her body. Now they crawl out, continue to develop outside into parasitic wasps, and then seek their sustenance from flowers and so forth. And then, when they mature, they lay their own eggs in turn into caterpillars such as this.

Well, there is something incredibly clever about this! Observing these things we can feel increasing wonder at how this can be, what is at work here, the whole context and background to it. Let's take a closer look at what's happening here. First, we have flowers growing forth from the earth. And then here are the caterpillars. Now along come these insects, feed their fill on flowers and caterpillars and keep on multiplying. The whole thing keeps starting afresh. And it may seem to us as if we could well do without this whole world of insects—except for the bees of course, for they supply us with honey, so apiculture seems useful to us. But this is all from the human angle. It is highly beneficial for us that the bees take nectar from the flowers, and we use this in turn, as honey, as food or even as a remedy. But from the point of view of the flowers this might appear mere thievery, in which we humans participate. Is this actually the point of view of the flowers—that these creatures, bees, wasps and ants, steal their nectar, and they would do much better if this were not so?

Human beings might assume so; and you may often hear people who know no better lamenting such natural 'misfortunes'—the poor flowers and the poor caterpillars whose lifeblood, as it were, is stolen from them. But this is not true, not in the least. If an insect, say a bee or some other insect, did not sit on the blossoms of flowers and trees and suck out the nectar, what would happen? This all becomes more complex than mere thievery, since we have to look into the whole balance of nature. In fact we cannot form a view at all about these things without looking back to former conditions of Earth evolution.

The earth was not always as it is today. Nothing could actually exist at all if the earth had always been as it appears to us today, an outer environment of dead chalk, dead quartz, dead gneiss, mica schist and so on, plus the plants that grow from today's seeds, the animals and so forth. None of this could exist if the earth had always been like this. People who base their science only on what is present today fall prey to complete illusion, since none of it could actually exist as such. Trying to fathom the laws of the earth from what science recognizes nowadays is like an inhabitant of Mars coming down to the earth without any understanding of living human beings, then looking inside a burial vault and seeing the corpses there. These dead could not of course be there if they had not first been alive! The Martian who had never seen living human beings would have to be educated about their existence—would have to be shown them. Then it would dawn on him that what he had first seen were the corpses of the living, who had first been alive and then died. Similarly, if we wish to understand the laws of earth's evolution, we have to return to earlier conditions. Our present earth was preceded by a quite different configuration. I have always called this the 'Moon stage'—that's how I refer to it in my book *Occult Science, An Outline*[64]—since today's moon is a residue of this ancient earth. And likewise this Moon stage was preceded by other conditions. The earth has changed and transformed and originally was quite different in nature.

On our earth there was once a condition in which no plants and insects yet existed as we now know them. Instead there was a body that can be compared with our present earth and from it grew plant-type forms, but ones that continually transform, that continually acquire other forms, like the clouds. These 'clouds' existed in the vicinity of what was then the earth, but they were not clouds like our present external ones, that are or at least appear dead, inanimate, but these were living clouds, with a life like that of our plants today. If you can imagine our clouds today coming to life and turning green, then you have a rough idea of the nature of the plant world in those times.

In this respect some scientists are very comical. Recently a ludicrous announcement was published in the newspapers about a new scientific discovery, very much in the modern style. This was very funny! It had turned out that milk, prepared in a certain way, was a good remedy against scurvy, a very serious disease.

What do scientists do nowadays? As I have said before, they analyse things, in this case milk. They discover that it contains certain chemical constituents. Yet as I have also said before, mice die after a couple of days if fed only with these chemical constituents, whereas they survive on milk itself. Pupils of Professor Bunge ascertained this fact and then simply concluded that milk, and honey too, contain a 'vital substance' above and beyond the analysed constituents—and they called this 'vitamin'! As I have also said before this is tautological—like saying that poverty comes from penury.

Now this important discovery was made: milk contains all kinds of substances with very scientific names. And, prepared in a particular way, milk can counteract scurvy. But in very scholarly fashion the scientists set to work to see if these things with scientific names contained in milk would cure a person with scurvy when given separately. Well, all these constituents proved ineffective! A person is only cured of scurvy when given milk itself, prepared in a certain way. The separate constituents do not cure the condition, only the whole substance of milk does.

So what remains, say the scientists, if we discount all these constituents? They do not acknowledge the fact that these constituents exist within an etheric body. They discount everything, and what remains? It must be vitamins! Thus it is thought that the vitamin that cures scurvy is not part of the analysed constituents of milk. Where is it therefore? And now they come up with the fine idea that it must be in the water content of milk, since it cannot be found anywhere else. And therefore what cures scurvy is water!

This is all very amusing, and yet it is the approach adopted by modern scientists. And if water contains the vitamin they are seeking, then we're pretty much at the point when scholars will see the clouds out there as animate, for looking at water vapour in the atmosphere we'd have to say there are vitamins in it everywhere. And then we would be back with a former condition of the earth. Except that today this is no longer so.

But in ancient times there was a plant unity, if you like, a living mantle of plants. And this living plant mantle was fertilized everywhere from what surrounded it. As yet there were no fixed animal forms: rather than wasps, only an animate, animal substance came from the wider environment. Thus our earth was once in a condition that we can more or less describe as surrounded with clouds that contained plant life, while other clouds, animal-like in nature, approached the former from without and fertilized them. This animality came from the cosmos, while the plant forms grew up from the earth.

All this changed. The plants of those times became our clearly defined flowers growing forth from the earth and no longer forming great clouds. Yet these flowers retain the desire to receive influences from their wider environment. Imagine a rose growing from the soil: here is a rose leaf, there another, a third and so forth. And now along comes a wasp that gnaws a piece from the rose leaf, carries it back to its nest and either uses it to build the nest or to feed its young. The wasp gnaws it off and bears it away. Now as I said, our rose bushes are no longer clouds but have become clearly defined and delimited. But what lived within such cloud forms, and was connected with the animality that came towards them everywhere, still remains in the rose leaves and rose blossoms. It still resides there. In every rose leaf is something that must inevitably be fertilized, in a sense, by its whole environment.

And you see, what these flowers need, what is essential to them, is a substance that also plays a major role in the human body. If you study the human body, you find it contains the most diverse substances. All these substances transform continually. But everywhere in the human body they are finally converted into something contained there in certain quantities, into formic acid.

If you find an ant on an anthill and squeeze it, a fluid comes out that contains formic acid and a little alcohol. This juice is inside the ants, and also finely distributed through their body. Whatever you eat is always converted into small quantities of formic acid among other things. It is present throughout your body. And if you are ill and do not have enough formic acid in you this is very bad for the body, which will develop gout-type or rheumatic conditions—and now I am coming to Mr Müller's question. Your body then forms too much uric acid and not enough formic acid.

In other words, ants contain something that the human body needs too. In fact formic acid is needed everywhere in nature. You cannot find even a piece of tree bark that doesn't contain a little formic acid. There is formic acid throughout the tree, just as in the human body. Every leaf has to have formic acid in it and, besides this, something else related to it that the wasps contain, the bees too, which then becomes their sting: bee poison. These insects all bear within them a certain poisonous substance. If a bee stings you, the area becomes inflamed. The same is true, although it can be much worse, if a wasp stings you. The business with wasp stings can be frightful. Brehm recounts a story about these insects and what havoc they wreaked on men and animals.[65]

It was like this. A cowherd, still a young fellow, was grazing cattle in the meadows, which were full of wasp nests. His dog was running about and suddenly went crazy, rushing about as if possessed, and no one knew what the matter was. He ran to the stream and threw himself into it, then shook himself violently and repeatedly. The cowherd was greatly alarmed, came running to the dog's aid and tried to help him without jumping in himself. But unfortunately, while standing there he trod on an insect's nest, as the dog no doubt had done, and he too was stung. Now he too ran about like a madman and finally also leapt into the stream. But because both the dog and the boy had escaped, a tumult now began amongst the herd of cattle. The cows that had trodden on an insect nest were likewise stung and went crazy. And eventually a large number of cows leapt into the water.

These insect stings can be a nasty business. All these creatures bear poison inside them; you'll even get a small inflammation from the formic acid that flows into the wound if an ant bites you. But this formic acid is, in turn, in all living things in the right dilution.

Now if there were no ants, bees and wasps to prepare these toxins, what would happen? The effect on human reproduction would be the same as if all men on earth were beheaded, leaving only women on earth. Human beings could no longer reproduce in the absence of male seed. These insects all have seeds as well, but despite this they need for their survival what comes from the toxins they produce—which are a residue from what once lived in the environment of Old Moon. Finely distributed bee poison, wasp poison, ant poison once entered plants from the cosmos, and a residue of this is still present today. So if you see a bee sitting on a willow tree or a flower somewhere, do not think that the insect is simply involved in thieving. As the bee sits there and sucks, the flower feels so good that it lets nectar flow towards the place where the bee is sucking. This is very interesting! As the bee sucks, the flower releases the nectar, sends it in that direction. And while the bee takes some of this nectar away, a little of the bee's poison is given in exchange and flows back. The same happens when the [gall] wasp stings and wasp poison flows in. And likewise especially as the ant makes its way over tree stumps and so forth, across things that are no longer alive, formic acid is introduced into them. When an ant comes along, the sap of the flower connects with the formic acid. This is necessary—and if it didn't happen, if there were no bees, wasps and ants connecting continually with the world of flowers, nibbling and sucking at them, the necessary formic acid and toxins would not flow into these flowers, and eventually they would die out.

You see, people value the substances they usually regard as sustaining life. But really only substances such as formic acid are life-sustaining. If you look at deadly nightshade, it contains poison, a harmful substance. But what is deadly nightshade actually doing? It is gathering to itself spirit from the cosmic environment. The toxins are spirit gatherers, and this is also why poisons are medicines too. Basically, the flowers keep getting more and more ill, and these bees, wasps and ants are little physicians that bring them the formic acid they need to cure their illness. So you can see that these bees, wasps and ants are not just thieves but at the same time bring the flowers what enables them to live.

The same is true of caterpillars: they would die out after a while, become extinct. Well, you may say that this wouldn't be too much of a loss. But birds feed on the caterpillars in turn, and so on. The whole of nature is interconnected. And when we realize that, for instance, ants pervade everything with their formic acid, we get a glimpse into the balance of nature. This is something magnificent. Everywhere something happens that is absolutely necessary for preserving and sustaining life.

Consider a tree, with its bark. If I fell the tree the bark rots away, producing bark mould. People just let this happen—everything in the wood that falls from a living tree decays and turns to leaf mould and suchlike. People just let it happen. But the rest of the world does not turn a blind eye to such things. In the wood there are anthills everywhere, and formic acid emanates from them into the forest floor.

Imagine the forest floor here, and here an anthill. It is like having a glass full of water to which you add a drop of something that spreads through the water and pervades it. If you add a pinch of salt, the whole water becomes a little salty. And if you have an anthill, the formic acid filters through into the whole forest floor, into the leaf mould, and the whole decaying forest floor is permeated by the formic acid. You see, the formic acid does not only pass into living plants and caterpillars—as does bee or wasp venom when the bee sits on a flower and the flower sucks up what it gets from the bee—but also into the decaying ground. Physical science is concerned only with what bees take from the flowers; but bees could not visit flowers for millennia if they did not in turn breed and rear them as they suck the nectar from them.

And the same is true even with the lifeless matter on the forest floor. Physical science today assumes that the earth will eventually become completely dead. This would indeed happen, for a condition would eventually prevail in which decay gained the upper hand and the earth died away. But this will not happen because wherever the earth decays it is at the same time interfused with what the bees, wasps and ants give back to it. Bees only give to living flowers, though, as do wasps, largely. But the ants give their formic acid to dead and decaying matter, and thus to a certain degree they kindle life in it and help ensure that the earth with its decaying matter remains alive.

So we can marvel at the spirit at work in all this; and when we look more closely at what is happening we see it to be of great significance.

Let us consider again these agricultural ants cultivating their little fields to produce quite different plants. A person could not use these grasses as food. These hard little rice grains, tough as flint, would firstly make him ill from assimilating too much formic acid. But besides that, he would break his teeth, and dentists would have plenty to do. And if a person persisted in eating these flinty rice grains, this would eventually lead to his miserable demise.

But the ants, the anthill itself really, says this to itself: If we sally forth into nature and suck from plants the substance that is present everywhere, we will obtain too little formic acid, and then we will be unable in turn to give the earth enough of the latter. Let us only choose plants that we can cultivate to render them very dense, hard as stone, and so we can get a great deal of formic acid from this compact density. And this enables the ants, in turn, to introduce much formic acid into the soil.

From this you can see that toxins with an inflammatory or similar effect are at the same time an enduring remedy for decay. And in this respect the bee, particularly, is of huge importance to ensure that the life of flowers is sustained, for there is a deep affinity between the bees and the flowers.

And this shows that whenever the insects engage with the earth like this, it is, if you like, replenished with these necessary toxins. That is the spiritual correspondence. Whenever anyone asks me about spiritual relationships I do not care at all to say they are like this or that; I prefer to give the facts and from these you can judge the meaning yourself. The facts themselves reveal the meaning present everywhere. Those who deem themselves scientists and academics do not tell you such things, but in life this plays a certain role. In our regions, less respect is accorded to such things but if you go further south you can hear farmers say, with more of an instinctive wisdom, that one ought not to destroy an anthill, for it helps mitigate decay. And those with a really sharp nose for these things—they are clever indeed, though not so much up top as in their noses—will tell you something more. If you go walking with them in the wood, particularly a forest where trees have been felled and a young plantation of trees is starting to grow, they will point you out a particular place and say that things will grow well there, will thrive, for there is less of a smell of decay than usual, and there must be an anthill nearby. These people can smell such things. This olfactory intelligence is the origin of much useful folk wisdom.

Unfortunately our modern civilization tends to cultivate the brain more, and has allowed such instincts to fade. And in consequence 'instinct' has become an empty word. Animals in their colonies, as hive or as anthill, fundamentally know all these things, through a kind of olfactory sense. And, as I said, the nose's intelligence is involved in much instinctive wisdom.

We will continue the session next week. Today I just wanted to say that the bees, wasps and ants not only take or steal something from nature but also enable it to go on living and flourishing.[66]

Discussion of the social insects leads us on to the next chapter. Rudolf Steiner compares an individual insect with the whole life of the colony and points out that at this level the single insect has little intelligence as such. Wisdom lives instead in the insect community. Steiner takes issue with ascribing all group behaviour simply to the instinct of each separate insect. His comments relate back to Chapter 1 and the question about whether the super-organism is composed of separate individuals or constituted by a superordinate whole. The visible flexibility of colony behaviour exceeds the sum of possibilities of all the individuals composing it. It is doubtful whether their coordination and collaboration can be fully explained by the extraordinary learning capacity of each insect, together with outstanding communication between them. The contrary view, that a single 'being' or entity (the super-organism) works through the social insects, is at least equally persuasive. This spiritual unity, which we could call Bee, Ant or Wasp, is regarded by Steiner as a group soul that uses the separate insects as its cells and organs.

'... so it is very hard to say that the insect simply has a generalized instinct, which would mean that everything it did was generally dictated by this instinct. No, the important thing is that the insect accommodates itself to outer circumstances.'[67]

A fundamental quality of a bee colony is its adaptability to outer conditions and the very diverse range of possible behaviours. A colony can establish itself in a space between two floors only 20 cm high, constructing many small combs on its surfaces, or it can build combs of two to three metres in length in a hollow tree. Besides this adaptability in relation to its dwelling place, the colony displays wide variations in behaviour, which are fascinating to beekeepers. It is astonishing to find things happening that one has never observed before even if one has been beekeeping for 20 years. This diversity of behaviour poses problems for a scientist since it signifies a very wide statistical spread and makes it hard to conduct conclusive trials on colonies. Hans Wille interprets this multiplicity of behaviours as a survival strategy of the colony.[68] It keeps trying different things in creative, playful fashion, giving rise to changing behaviour that makes survival possible even in the most difficult conditions. This wide range of behaviour goes a long way to relativizing the concept of instinct.

A key theme in this lecture, however, is the relationship between insect and plant worlds, based on reciprocal giving and taking in which the plants receive from insects something that is not immediately apparent to us: 'So you can see that these bees, wasps and ants are not just thieves but at the same time bring the flowers what enables them to live.'[69]

The prevailing view nowadays of the relationship between plants and insects is different from this:

The green plants have the capacity to synthesize simple

organic compounds from air, water and sunlight. This photosynthesis underlies all life, building up organic life from inorganic elements. Incorporation of minerals and nitrogen gives rise to more complex compounds and more solid substances. Plants are thus producers in the biosphere, as opposed to the consumers—animals and humans. Their life depends on breaking down and converting organic substance, and thus they rely on the prior work of the producer-plants. This breakdown and assimilation is connected with the process of respiration. For the sake of completeness, we should also mention here the decomposers, mostly fungi and bacteria, which break down organic compounds into their mineral constituents again.

Insects account for 80 per cent of all animal life. They are a group very rich in species, accomplishing the most diverse range of catabolic processes. Inhabiting the most varied 'corners' of our world, they are always present in close association with plant life. They include species that can undertake a wide range of decomposition tasks, while others are very specialized. Typical specialists include the bee or wax moth, the only creatures that can break down wax.

Bees, by contrast, stand at the very outset of decomposition processes: pollination of flowers usually sets plant decay in motion since after this happens they concentrate on the involutionary process involved in fruiting and seed formation. Wasps and ants, on the other hand, are occupied with a wide spectrum of decompositional activities on dead plants and animals.

All insects have the mission of breaking down substance to create space for new life. The body of the once living being is broken down into its constituent substances. The better this breakdown occurs the faster new life can arise. People frequently regard insects as pests since they seem to be working counter to human intentions, causing damage to wood, food stores and cultivated plants. They are also seen as disgusting or unpleasant—an insect landing on you is likely to elicit a defensive reaction, or at least a shiver.

'Thus our earth was once in a condition that we can more or less describe as surrounded with clouds that contained plant life, while other clouds, animal-like in nature approached the former from without and fertilized them. This animality came from the cosmos, while the plant forms grew up from the earth.'[70]

This striking picture gives a sense of the interplay between plant world and animal world—an overall panorama that avoids us getting lost in the numerous details of reality. The encounter of cloud and cloud encompasses the most diverse interactions uniting the animal and plant realms; and this simplified picture conveys the qualities involved in the movements that arise in this encounter. What strikes us here is the intimate nature and necessity of the plant-animal relationship. According to Steiner, the plants depend on receiving formic acid from insects, both through their direct contact and also through the air. This formic acid enables the plants to be continually revived and renewed.

In modern journalism, the bees' pollination services are calculated as a value in euros, whereas the breakdown labour of insects is summed up as damage and debit. While others have highlighted the intimate affinity between insects and plants, only Rudolf Steiner speaks of the former healing and reinvigorating the latter:

So you see, we can contemplate two sentences that express a great secret of nature:

Regard the plant—
a butterfly
fettered by the earth.

Regard the butterfly—
a plant
liberated by the cosmos.

The plant—the butterfly chained by the earth! The butterfly—the plant liberated from the earth by the cosmos!

Consider the butterfly, or any insect in fact, from egg stage to the fluttering insect: it is the plant raised into the air, configured in the air by the cosmos. Then consider the plant: it is the butterfly that is fettered below.[71]

Elemental beings

Alongside these accounts of mutual dependency and inner affinity, Rudolf Steiner also describes a third phenomenon, that of the elemental beings that appear in the dynamic sphere between bees and blossoms. It is fascinating to watch all the insects visiting a flower, and creates a particular mood in us. Wakefulness to such moods and experiences can open new spheres of perception.

People today speak a lot about natural forces, but very little about beings that underlie these natural forces. To speak of nature spirits today strikes modern people as a mere revival of old superstitions. The names our forefathers used for these spirits, such as gnomes, undines, sylphs and salamanders, are not thought to have any reality. In some respects it is unimportant what theories and ideas people have; but when these theories blind them to certain things and govern the way they act in actual life, the matter starts to be a great deal more important.

Who believes nowadays in beings whose existence is

bound up with the air, or who are embodied in water? When someone ridicules old beliefs about gnomes, undines, sylphs and salamanders, and dismisses them as fantasy, one feels like saying: 'Go and ask the bees.' If the bees could speak they would say: 'The sylphs are not superstition, for we well know what they give us!' And someone whose spiritual eyes are opened can observe what power draws a bee to the flower. The 'instinct' and 'natural drive' posited by human beings are empty words. Living beings guide bees to the calyx to seek food there; and in the whole swarm of bees that set out to seek sustenance, beings are at work that our forefathers called sylphs.

Wherever different realms of nature meet, there is an opportunity for certain beings to manifest. Particular beings engage within the earth, for instance, where stone meets metal. Or by a spring where moss covers stone, so that the plant and mineral kingdoms meet, such beings settle. Where animal and plant meet, in the blossom, where the bee meets the flower, certain beings are embodied—and likewise where the animal and human realms encounter each other. This is not the case in the ordinary course of things—not, say, when a butcher slaughters a cow or when someone eats meat, not in such everyday occurrences. But rather such beings come to embodiment where, as between bee and flower, two realms touch each other as if in a superfluity of life. And especially this happens where human sensibility and intelligence is especially engaged with animals in a mutual relationship such as that between a shepherd and his sheep. In a feeling connection of this kind, such spirits come into being.[72]

9. PROCESSES

We still need to speak a little about Mr Dollinger's question. On your behalf—for this is no doubt of interest to all—he was enquiring about the spiritual relationship between the swarms of insects that alight on plants and what lives in the plants themselves.

I have said this before: not only oxygen and nitrogen surround us everywhere but there is intelligence, real intelligence present in all of nature. No one is surprised to hear that we breathe in air, since it is everywhere. And scientific insight that air is all around us and we breathe it in has found its way into all the school books and become common knowledge. Yet I have known people in rural areas who thought this was fantasy since they had no idea there was air around us, just as people today have no idea that intelligence exists everywhere too. They regard it as fantasy if you tell them that just as we breathe in the air with our lungs, so we inhale intelligence through the nose or ear. In the past I have given numerous examples of the omnipresence of intelligence. Recently we have been speaking of a particularly interesting field of science, about bees, wasps and ants. Actually the life of insects may give us deeper insights into nature itself than almost anything else. The insects are very remarkable creatures, and there are still many secrets of nature that they can reveal to us.

It is noteworthy that we are discussing insect-related matters close to the centenary, on 22 December, of the important insect researcher Jean-Henri Fabre.[73] Though he lived in a fully materialistic age and therefore interpreted everything materialistically, he discovered a huge number of facts about the life of insects. And so it is fitting that we remember him today, when speaking about insects.

Now I'd first like to offer the example of an insect species that can be of great interest particularly in connection with bees. The work of bees is close to perfection; and the most remarkable thing about them is not the honey they produce but the wonderful combs they create out of themselves. The material they use they must first carry into the hive themselves, and rather than using it in its original form they transform it entirely; and thus they work out of themselves.

But there is a type of bee that does not work like this, and whose work clearly demonstrates what huge intelligence lives in all nature. Let us consider the work of this species of bee, usually called the carpenter bee[74]—a creature not as well-regarded as the honey bee since it is usually a nuisance for human beings. This is a very industrious insect that has to accomplish a huge amount of work to survive—I mean for the whole species to survive, not the single insect. It seeks out wood that is no longer growing but has been felled, and works on it. For instance, you can find this carpenter bee with its nests—which I will describe to you in a moment—in fence posts or wooden pillars, wood that is no longer growing. The carpenter bee can also be found in garden benches or doors—wherever people use wood, the carpenter bee will make her nest in a very remarkable way.

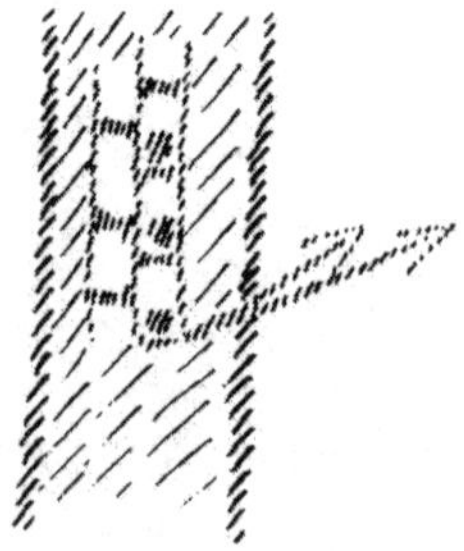

Imagine a fence post.

So this is wood that has been taken from a tree. The carpenter bee comes along and from outside bores a slanting passage into the wood. Arriving inside, having excavated her passageway, she starts boring in a quite different direction, carving out a small, circular cavity. Then the insect flies away and from the surrounding area fetches all sorts of materials to pad out her nest. Having done this she lays her egg there, which will turn into a larva. After laying the egg the bee lays a cover or lid over it, with a hole in the middle. And now she starts boring further above this lid, creating a second cavity for another egg to hatch in, and likewise padding this, and placing a lid over it with a hole. The carpenter bee carries on doing this until she has built ten or twelve nests one above each other, each one containing an egg.

Now the larva can develop inside this wooden post. The bee places food beside each egg for the grub when it hatches. First it eats this and then grows ready to pupate; and then it transforms into the winged bee that can fly away.

So inside the post the grub develops and after some time pupates, then hatches as a bee that emerges through the original passageway. All well and good. But the second insect, that hatches out later and is a little younger, and then the third, younger still, have no side entrance to fly out through. The insects hatching out higher up the post would by rights die in their cavity. But the mother bee has prevented this by laying her egg so that when the younger grub hatches she finds the hole I told you about and lets itself fall through it, and can thus emerge. Likewise the third insect in the hierarchy lets itself fall down through two holes and emerge. By virtue of the fact that each insect hatches out later than the one before, it does not disturb the one below it which has already hatched and flown the nest. They never end up together; the older one has always flown away by the time the younger one is ready to do so.

The intelligence at work here is astonishing. The machines people make nowadays in imitation of nature are usually much less skilfully done. You have to acknowledge that there is intelligence at work here, real intelligence. One could cite hundreds and thousands of examples of the intelligence at work in the way insects build and labour. Remember the agricultural ants and their whole cultivation work and you will see how much intelligence is at play here.

Now we considered something else too when we were talking about these insects, bees, wasps and ants. I told you they have within them a kind of poisonous substance; and this toxin, which all these creatures contain, is at the same time also an excellent medicine in the right dosage. Bee venom is an excellent remedy. Wasp venom too. And formic acid, which ants secrete, is especially beneficial. As I said, if you pick up an ant and squeeze it, formic acid is exuded. So the ants contain it within them, and by squeezing one we can obtain it. This formic acid is found primarily in ants. But you would be astonished if you knew how much formic acid, relatively speaking, is present here among us now in this hall. When I say this you might turn round to look in the corner and see if there's an ants' nest here. But no. All of you sitting here are in reality yourselves an anthill! There is formic acid everywhere in you: in your limbs, muscles, other tissues, in the heart tissue, lung tissue, liver tissue, pancreatic tissue; not in such concentrated form or as strong as in the anthill, but nevertheless it thoroughly permeates your body. That is a very curious fact.

Why do we have this formic acid in our body? If someone has too little of it, we have to recognize this. If someone displays an illness of some kind—and people are usually a little bit ill—he may have any one of hundreds of conditions that all seem the same outwardly. One has to discern what is actually wrong—pallor or lack of appetite are only outward symptoms. We have to see what is actually wrong. In some cases a person may simply have too little of an anthill in him, is producing too little formic acid. Just as formic acid is produced in the anthill, it must be vigorously produced too in the human body, in all its limbs, especially the pancreas. And when a person produces too little formic acid, you have to give him a preparation, a medicine, that will outwardly help him in creating sufficient formic acid.

But now you have to observe what happens to someone who has too little formic acid. And such observations can only be made by those who have a good knowledge of human nature. You have to form an idea about what is happening in the soul of someone who first had enough formic acid in himself and later on too little. This is very curious. If you question him in the right way, such a person will tell you the truth about his illness. Imagine that someone says in reply to your questions, 'Oh goodness, a few months back my thoughts were all in order, I was able to work things out and connect one thought to another. But that's no longer so. It's not working any more. When I want to reflect on something, I can't do it.' This is often a much more important indicator than all external tests, although these must of course also be done nowadays. Today you can test the urine for protein, purulence, sugar and so on. Naturally this provides very interesting results. But in some circumstances it can be much more important that a person recounts what I just described. If he tells you something like that, you will still need to find out a few other things of course, but you can discover that the levels of formic acid have declined too much in his body in recent weeks.

But now someone who thinks in overly external ways may say, 'OK, he has too little formic acid in him. I will give him a corresponding dose of formic acid.' You can do this for a while and then the patient will come to you and say, 'This hasn't helped at all.' What has happened? This really hasn't helped him. It was quite true that he had too little formic acid in his system, and therefore you gave him formic acid, and yet this was of no use whatsoever. Why is that?

If you enquire further you will find that formic acid brings no benefit to one person but helps another considerably. Gradually you can discern the difference. Those people whom formic acid benefits will display mucous obstruction in the lungs. Those who gain no benefit from formic acid show mucal obstructions in the liver, kidney or pancreas. This is a curious matter. There is a great difference, therefore, whether the lungs lack formic acid or the liver. The difference is that the lungs can accomplish something with this formic acid, which is present too in the anthill, whereas the liver is unable to do anything with it.

And now something else enters the picture. If you notice that someone has a disorder of some kind in the liver or especially in the intestines, and formic acid does not help him despite the fact that he is lacking in it, you have to give him oxalic acid. You must take ordinary sorrel or clover in general as you find it in the fields, press it to extract the acid and give this to the patient. Thus someone with a lung disorder needs formic acid whereas another person with a liver or intestinal disorder needs oxalic acid; and eventually he himself will convert this within himself into formic acid! So it is not just a question of introducing into a person what is lacking, but you also have to know what the organism itself accomplishes independently. If you give 'ready-made' formic acid to a person [with a liver disorder], his organism says: 'This isn't for me; I need to work. This gives me nothing to work on; I can't get this up into the lungs.' Naturally you have to introduce this into the stomach, and from there it eventually reaches the intestines. The organism, wishing to work, asks, 'What is this that I'm being offered? I am not meant to create the formic acid myself but I'm meant to get the formic acid I am offered up into the lungs. I'm not doing that.' Instead it wants oxalic acid, and from this it makes its own formic acid.

Yes indeed, life involves work not substances, and it is of the very greatest importance to know that life does not consist in consuming cabbage and carrots but in what the body has to do when the cabbage and carrot substances enter it. At least it ought not to fabricate a cabbage head again from the cabbage it ingests. But curiously this is the principle underlying our modern civilization.

From this you can see the remarkable interconnections at work in nature. Outside us are the plants. Wood sorrel characteristically contains oxalic acid but so do all other plants. Wood sorrel contains the most, that's all, hence the Latin name for wood sorrel, *Oxalis*. But just as there is formic acid everywhere in nature, and throughout the human body, so this is also true of oxalic acid.

Now here's an interesting fact. Take a test tube, light a flame underneath it and now put oxalic acid into it—in the form of a powdery, salty ash; then add an equal amount of glycerine. Mix them together and then heat them. Then I get these vapours [see drawing]. I can capture this vapour but at the same time I notice that air is expelled everywhere. If I test this air I find that it is carbon dioxide. And here, where I capture what condenses, I get formic acid. This contains formic acid. In the test tube I put oxalic acid and glycerine.

The glycerine remains there, while the other substance vaporizes and condenses in droplets here below as formic acid, and here the carbon dioxide is expelled.

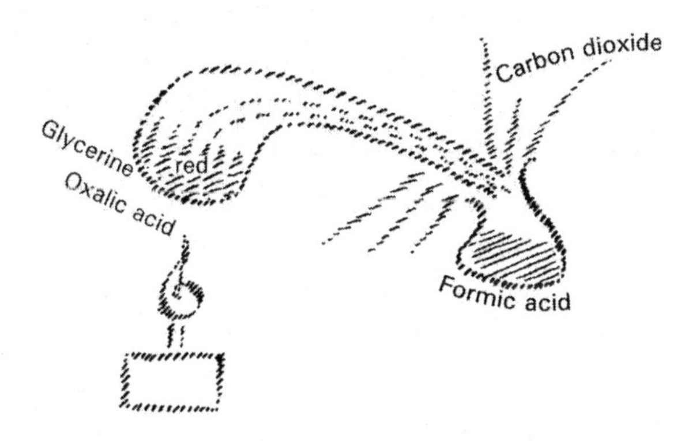

Now, let us we consider this again in different terms, replacing the test tube with the human liver, or some human or animal tissue, some organ of the animal abdomen—liver, pancreas or suchlike. I introduce oxalic acid through the stomach. The body itself already possesses the glycerine. In my intestines I then have oxalic acid and glycerine together. And what happens? Carbon dioxide comes out through the mouth, and from the lungs formic acid drips down everywhere into the organs. So the whole thing I have drawn here is what we have in our body. In our body we continually synthesize formic acid from oxalic acid.

Now picture the plants spreading everywhere over the earth: all of them contain oxalic acid. And now recall the insects. This manifests here in the most remarkable way. Think first of the ants: they come to the plants, or also to the decaying remains of plants. All of this contains oxalic acid; and from it they synthesize formic acid just like the human being, thereby ensuring that this formic acid is everywhere present.

A philistine looks at the air and says it contains nitrogen, oxygen. But insects whirr through the air too, and in consequence it contains very small quantities of formic acid as well. So we have, on the one hand, the human being, a microcosm, who makes formic acid and continually imbues his breath with it. And out in the greater world, the macrocosm, instead of what occurs within us we have the hosts of the insects. The great breath of the air enveloping the earth is continually pervaded by formic acid which is synthesized from the oxalic acid in plants.

If we observe correctly and study the human abdomen with the intestines inside it, the stomach, liver, kidney, pancreas and so on, we find that oxalic acid is continually being converted there into formic acid, and, through the air that we breathe in, this formic acid passes into all parts of our body. That's how it is in us.

Outside on the earth you have the plants everywhere, then the great diversity of insects fluttering over them. Here below you have the oxalic acid. The insects flutter to the plants and through this encounter formic acid is created and fills the air. And so we also always inhale formic acid from the atmosphere. Wasp poison is a venom related to formic acid, just somewhat transformed. And likewise the venom bees have in their sting—although in fact it exists throughout their body—is again converted formic acid, a more refined form of it. Taking this into consideration, we look at these insects, the ants, wasps and bees, and find that what they are doing is extremely clever. Why so? If the ant had no formic acid, all those fine forms of behaviour I described would be accomplished instead very dully. It is only because ants can create formic acid that everything they build appears so intelligent and wise. The same is true of the wasps and the bees.

Don't we have good cause, therefore, if we ourselves produce formic acid inside us, to recognize that there is intelligence everywhere in nature around us, and this arises through formic acid? Within us intelligence is also at work everywhere, because we too have formic acid. And this formic acid would not exist if the oxalic acid were not there to begin with. Now the insects flutter and hover over the plants, and cause the oxalic acid in these plants to be transformed into formic acid, to undergo metamorphosis.

We only understand these things if we first enquire into the nature of oxalic acid. Oxalic acid exists wherever life is to unfold. Where something lives, there is oxalic acid. But an etheric body exists there too. The etheric body ensures that the oxalic acid is immediately renewed. Yet this oxalic acid can never become the kind of formic acid which the human or animal organism can use if an astral body does not convert it. You see, the formic acid that I have taken here from the test tube is of no benefit to the human and animal body. It is dead, and it would be wrong to think that it is of any real benefit. The formic acid engendered here and here—within the human being and by insects—is alive and appears wherever sentience and soul arises. The human being has to develop formic acid in himself if he desires to develop soul and sentience from the mere life that exists in his abdomen, where oxalic acid plays an important role. Then this soul element lives in the formic acid in the breath, rising upward to the head where it can continue to act. The soul needs this processing of oxalic acid into formic acid within us.

What actually happens when oxalic acid is transformed into formic acid? What we spoke of first today can teach us this. This carpenter bee I spoke of is particularly interesting, since she bores her way into wood that is no longer alive. If this carpenter bee could not make proper use of the wood, she would make her nest elsewhere. This bee does not make her nest in trees, but in decaying wood, the wood of fence posts, say, that is already starting to decay. She first bores her way in, creates a nest inside them, then lays her eggs there.

If we study the connection of decaying wood with the carpenter bee, we discover that what happens here inside the decaying wood is the same as occurs continually within our body. The human body starts to decay, and if it decays too far it dies. And what happens out in nature is something we too must continually do: we must build up our cells—something that is only possible by converting plant substance, imbued with oxalic acid, into substance pervaded by formic acid.

So now you may ask what all this signifies for nature. Let's imagine a wooden fence post that is starting to decay. If a carpenter bee should never bore into it, farmers would be delighted, since it will be all hollowed out, weakened critically and a new one will soon be needed. People will not be happy about this, but nature is all the happier. You see, if all the wood that originates in the plant kingdom continued to exist without these bee nests, eventually the wood itself—as you can see from the way it decays—would crumble away and turn to dust; it would become quite unusable. But the wood in which a carpenter bee has been working does not pulverize but is instead enlivened again. And from all the wood that is a little bit enlivened again through the labours of the carpenter bee, there arises—and does so also through all the other insects—much of what will ensure that our whole earth, enlivened by the insects, can live on rather than decaying entirely and turning to dust in the cosmos. In nature, the formic acid which the insects produce from the plants' oxalic acid acts in a way that ensures that the earth can go on living.

Consider this whole context. Here we have human beings and here the earth. Look at the human being first. Let's picture a young child. When he is young he easily converts the oxalic acid in his abdomen into formic acid. His organs get enough formic acid. The human soul develops in the child. Thus we have this formic acid as the foundation for soul and spirit. And now, when the person grows old, and can no longer develop sufficient formic acid inside him, his soul and spirit depart. In other words, formic acid draws in soul and spirit; and if this were not to happen, the spirit would depart. This is very interesting.

A person suffering from much internal suppuration is helped to overcome these festering processes inside him if you give him formic acid. Then the right relationship between the astral body and his physical body—hindered by these processes—is re-established. Thus formic acid is always needed in the right way as foundation for soul and spirit. If the body has too little formic acid, it starts decaying, and can no longer keep the soul: the body grows old and the soul must depart.

Here we have the human being on the one hand and nature on the other. In nature, oxalic acid is continually converted into formic acid so that the earth is always able to be enveloped not only in oxygen and nitrogen but also formic acid.

This formic acid means that the earth does not, if you like, die every year but each year can be renewed and enlivened in this upper region. What is below the soil yearns as seed for the formic acid above. And the renewal consists in this. Each winter the spirit of the earth, we can say, endeavours to depart. And in spring the spirit of the earth renews and enlivens itself again. The spirit of the earth makes the earth grow rigid and frozen in winter; in spring it enlivens the earth again. This renewal occurs by virtue of the seed, which waits below the soil, growing towards the formic acid that was created the previous year through the interplay of the insect world with the plant world. Seeds not only emerge into oxygen, nitrogen and carbon, but also into formic acid. And this formic acid stimulates them to develop oxalic acid themselves in turn, by virtue of which the formic acid can be present again the following year. But just as formic acid in us can be the foundation for our soul and spirit, so the formic acid spread throughout the cosmos is the foundation for the soul and spirit of the earth. And so we can say that what holds true for us holds true for the earth as well: formic acid is the matrix for earth soul and earth spirit.

Did you know that it is actually harder to send telegraphs in a region where there are no anthills than in one where they are present, since electricity and magnetism, which are involved in telegraphy, depend on formic acid? When telegraph wires pass through cities where there are no ants, there must be enough power coming from outside the city, where the wires pass over fields, for the magnetic and electrical currents to get through. But of course formic acid spreads far and wide and also fills the air of cities.

What exists within us is also outside us in nature, and this applies to the generation of formic acid too. The human being is a little world, a microcosm. During our lifetime, until we die, we are organized so as to produce formic acid from oxalic acid. Then we become unable to do so, and the body dies. We have to acquire another body which, in childhood, can make formic acid from oxalic acid in the right way again. In nature things simply continue: winter, summer; winter, summer. Oxalic acid is continually transformed into formic acid.

If you observe a person who is dying, you can get the sense that he is testing whether his body is still suited to developing sufficient formic acid. And when the body is no longer suited to this function, death intervenes: the person enters the world of spirit and can no longer endure being in his body. And so we can say that a person dies at a particular moment; then a long time passes, and he enters a new body again. In between he tarries in the world of spirit.

Now when a new queen hatches in the hive, there is, as I told you, something that disturbs the bees. Previously they lived in a kind of twilight, and then they see this young queen shimmering. What is this luminosity of the new queen connected with? Her shimmering is connected with her depriving the old queen of the power of the bee venom. And that is the fear of the emerging swarm: that it no longer has its bee venom and so can no longer defend itself, save itself. And so it leaves. Just as the human soul departs at death when it can no longer have its formic acid, so the old bee community departs if there is not enough transformed formic acid—bee venom, in other words. If you observe a swarm of bees, although visible it does closely resemble the human soul that must depart from the body. A great swarm departing from the hive is a magnificent sight, an image of this. Just as the human soul leaves the body, so, once the young queen has hatched, the old queen leaves the beehive with her attendant throng; and in the departing swarm we can truly see an image of the human soul flying forth from its dwelling.

Oh, this is such a magnificent thing! The difference is only that the human soul has never developed its faculties as far as becoming many tiny creatures, though this tendency exists in us continually: we wish to become lots of little 'critters'. There is in us really an impetus to transform ourselves all the time into scuttling bacilli and bacteria, into many tiny bees, but we keep suppressing this tendency; and this makes us into whole people. The bees cannot find their way into the world of spirit; instead we must help them to reincarnate in a new hive—a precise image of the reincarnating human soul. Whoever can observe this will feel huge respect for this swarm of older bees with their queen, who really behave as they do because they seek entry to the world of spirit. But they have become so materially physical that they cannot. Instead the bees cosy up to each other, cohere and become a single body. They want to become one and depart from the world. As opposed to their usual flight they now throng together on a tree branch or suchlike, huddle and nestle together so as to vanish, because they want to enter the spiritual world. And then they become a proper colony again when we help them by returning them to a new hive.

Insects teach us about the most sublime reality of nature. In ancient times the instinctual vision of plants that people possessed enlightened them about something that I have described to you but which has been lost sight of completely in modern science. Such people looked at plants with special eyes. In our own times people get an inkling of this, a faint memory, when they bear a fir tree into the house and turn it into a Christmas tree. They are then reminded how something that exists outside in nature can become something which, brought inward into human life, plays a part in social interplay. This should be an image of love, this fir tree transformed into a Christmas tree. People often think that the Christmas tree is very ancient, but in fact it is a relatively recent tradition, only 150 or 200 years old. But prior to the introduction of this custom, people did use a kind of bush or shrub at Christmas, for instance in the Christmas plays that were performed in rural villages in the fifteenth and sixteenth centuries. In Britain, a simple evergreen tree, the 'paradise tree', featured in the plays. In central Germany someone holding a small juniper bunch in their hand would run around the village to announce the imminent performance. Why juniper? Because its berries, so beloved by birds, symbolized the mild toxic effect that must penetrate the earth so that spirit can emerge within the earthly realm. Just as when the ant harvests mouldering wood or the carpenter bee bores into a fence post, so each morning when birds gather to peck at the fruit of the juniper acid is produced everywhere, albeit in much weaker form. In olden times people knew this instinctively: that in winter, when the juniper tree bears its ripe berries, and the birds flock to eat them, the earth is renewed and rejuvenated by this. And for them this was also an image of the moral rejuvenation of the earth by Christ.

Properly observed, what happens outside in nature offers true images of what happens in human life. In olden times, when people saw birds sitting on a juniper tree, they regarded them with the same love that we feel today for little pasties and mince pies, or the presents under the Christmas tree. For these people the juniper tree out of doors was a kind of Christmas tree, and they brought branches of it into their dwellings and so made a kind of Christmas tree with it.

It is time to finish now. But—since at present you are under particular strain—I didn't want to let our session today end without touching on a very important theme: mentioning these little bunches and bushes of juniper that can really be seen as small Christmas trees, and give birds the same that bees gain from plants, ants and the carpenter bees from wood, and that insects in general garner from wood. I'd like to take this opportunity as we end to wish you a very happy, joyful Christmas festival that raises your spirits and lifts your souls.[75]

This lecture focuses particularly on the formic acid process. Chapter 4 discussed the silicic acid process, and also referred to the oxalic acid process and the carbonic acid (carbon dioxide) process. All four processes are regarded as indispensable for life, each in a different realm. Steiner considers the development, transformation or crystallization of organic acids in chemical and physical terms, describing their action both within the human being and in relation to the whole earth and its evolution—in line with the principle that the human being is an image and reflection of the cosmos.

To get inside these descriptions of acid processes, we have to find a different approach. If we first become conscious of our own perspective and view of the world, we can then compare it with Steiner's concerns and the way he proceeds.

Our way of thinking leads us to try to understand life in terms of its organic constituents. Biochemical processes form the basis upon which this organic life develops. The life processes of man and animal, as described, are based on breaking down plant and animal substance. Energy for the function of the organism is obtained from breaking down carbohydrates; protein is separated into its constituents and becomes available for resynthesis. Drawing on these two complex processes, people seek to explain the origins of life: the structure of an intact organism makes it possible, in this scenario, for a soul to develop somewhere within, as a dwelling for powers of will and reason. In our head we have the organs of perception, which can be seen as portals opening to the world, and supplying the brain with information. The brain, in turn, has a certain governing and guiding function. But we also see the functions of our soul and spirit as dependent on the organic foundation of our body. Yet from this point of view, studying the above-named processes—of silicic acid, formic acid, carbonic acid and, last but not least, oxalic acid—will not get us any further.

Rudolf Steiner proposes a different picture of the human being. He distinguishes four different realms: that of the soul and the spirit on the one hand, then of life processes and the physical domain on the other. Only the physical aspect is immediately apparent to us, while the others are supersensible or spiritual in nature. From the point of view of sense-based perception and science, the foundation of everything is the physical body, whereas Steiner starts from the spiritual, supersensible worlds.

This distinction is most vivid in different views of death. If we see the organic realm as primary, death comes at the moment the body loses its life functions and vigour. Rudolf Steiner offers us a different approach: 'Thus we have this formic acid as the foundation for soul and spirit. And now, when the person grows old, and can no longer develop sufficient formic acid inside him, his soul and spirit depart. In other words, formic acid draws in soul and spirit; and if this were not to happen, the spirit would depart.'[76] Accordingly, the dying process can be determined not only by the organic level but also by the soul. This opens up a fascinating perspective in relation to life and death, and also offers an outlook that starts from the spirit.

Here Steiner sees I organization and thinking as primary, and on this foundation seeks to pinpoint the process that best reflects this activity. For him, this is the silica process. Then he considers soul processes and enquires spiritually into their foundations, here outlining the formic acid and the carbonic acid processes. These appear, therefore, as the material counterpart to soul life. Finally he comes to the life forces and assigns these to the composition of oxalic acid.

Thus he proceeds in the reverse direction: instead of seeing DNA as primary, Steiner starts from the I organization, the etheric-life realm and the soul-astral realm. The processes he describes are to be understood as a reflection of the spirit within material occurrences. They can also be regarded as parallel phenomena within the physical domain. Just as all matter has a spiritual aspect, these spiritual processes have a material foundation. A description of them does not have to exactly reflect our reality since they are, rather, pictures of spiritual reality.

We are left with the question as to what these pictures can signify for us. Do we try to measure them against the biochemical reality of our world-view, or should we see them as analogies? No doubt their value lies somewhere between these. While we should compare them with our understanding of chemical processes, their focus lies in dynamic process rather than in base and end product. Their value encompasses picture and metaphor.

The silica process

We considered the silica process in Chapter 4; and here it is examined again in terms of the process picture. A physical process is described which continually seeks to form crystalline hexagonal structure from fluid silica. Viewing this pictorially, the silicic acid in the organism always has a tendency to become mineralized: thus it endeavours to evade life, instead choosing lifeless, mineral, crystalline form. We can understand or at least intuit that the I is related to perception and thinking, and that these activities are centred in the head and brain. According to Rudolf Steiner, the silicic acid process proceeds from the head, and is governed by the I organization; its realm is thinking.

Rudolf Steiner describes as follows the relationship of the I, the brain and the intellect to this lifeless mineral realm:

Mineral realm—here again there is a complex of forces that creates the diverse forms of the mineral kingdom. With the I this is very apparent, since you can after all only think the mineral realm. It has been endlessly stated that the intellect can only grasp what is dead. In other words, what exists in the I comprehends what is dead. And so our I lives in this complex of forces that forms the mineral kingdom.[77]

If we really consider this properly, we find a wonderful contrast in the human being. In our blood's white corpuscles we bear cells full of life; they wish to live continually. And in our brain we bear cells that seek really always to die, are always embarking on death. And this is true: through our brain we are always travelling the path of death; the brain is really always at risk of dying.[78]

In modern times we have intellectual experience. Inwardly, the intellect renders us cold, dead. The intellect paralyses us. We're not really alive when we develop our intellect. We just have to have a sense of this, that we're not really alive when we think, that we are pouring our life into dead, rational pictures, and that we need vigorous life if we are to feel that what exists in this dead rationality does, nevertheless, contain creative life [...].[79]

The formic acid process

The formic acid process is the major theme of this last section. Rudolf Steiner describes how formic acid is created in the laboratory. By heating oxalic acid and glycerine, we obtain formic acid, while carbon dioxide is given off. Steiner points to the material basis here, but immediately leaves this to one side again since he is concerned with dynamic life processes rather than base and end products.

Now the human being is not a test tube. The test tube shows us, in a dead fashion, what exists in us as life and feeling. But it is true to say that if we were never to form oxalic acid in our digestive tract we couldn't live—that is, our etheric body would have no organic foundation. Our etheric body needs oxalic acid, and our astral body formic acid. And in fact we do not need these substances as such but the work, the inner activity involving the oxalic acid process and the formic acid process.[80]

As modern science understands it, formic acid is one of the simplest organic substances. It exists on earth in small amounts in plants, animals and the human being, though never in a pure state. It is also present in concentrated form in many insect venoms.

In bound form, oxalic acid and formic acid play a role in our body. As part of a greater organic complex they belong to a key biochemical process that was discovered in the 1940s, and is known as the citric acid cycle or also the Krebs cycle (after Hans A. Krebs who discovered it).

In Rudolf Steiner's descriptions, oxalic acid and formic acid are however not just two important substances in metabolism, but also form the basis for the interplay of soul and body. The formic acid process stands for breakdown or catabolism, which in turn sustains soul life. This is because soul-spiritual processes, all thinking and emotional life, have a physically catabolic action. Thus the oxalic and formic acid processes represent the interaction between breakdown and synthesis, and in a sense also life and death.

Steiner says that the relationship between soul-spirit and bodily life is characteristic of human life; it is a relationship we can experience in the sleeping and waking rhythm. When we're awake, the soul connects with the body, while in sleep it detaches itself. Breathing belongs to the primary process of soul life. The body takes up oxygen, which spreads into the smallest cell structures. The totality of metabolism needs oxygen; and at the end of these processes, carbon dioxide and nitrogen are expelled via the lungs. The connection between breathing and soul activity can be observed in all kinds of situations. When we're tense or stressed, our breathing unconsciously becomes shallower, whereas relaxation or also meditation is distinguished by calm, deep breathing. When the soul loosens itself from the body in sleep, a quite different kind of breathing arises. If we ask where the seat of the soul is in the body, then breathing, the chest and perhaps also part of the abdomen suggest themselves immediately. But in the descriptions in the lecture, Steiner points to the formic acid process as the foundation of the soul, with the attendant task of renewing life forces. The organism cannot sustain its life by its own means: soul activity is needed to sustain organic processes. For this, Steiner employs the picture of formic acid, which 'drips down everywhere' into the human body.

Formic acid creates the ongoing impulse for renewal of life—an impulse that for instance facilitates cell renewal in the human body. Ultimately this is a soul impetus for life forces; and in summary we can say that this formic acid process is the soul's contribution to sustaining and preserving the unity of body and spirit.

Elsewhere, Steiner describes aspects of spiritual enquiry

and endeavour in terms of the alchemist and his experiences as he experiments:

All these true enquirers [. . .] feel one thing deeply: that when they experiment, nature spirits speak to them—the spirits of earth, of water, of fire, the spirits of air. [. . .] One has a test tube before one, and piously immerses oneself in what occurs. Precisely in this process in which we witness the metamorphosis of oxalic acid into formic acid, we can experience how [. . .] the spirit in nature replies to us, so that we can employ it for our inner being. The test tube thus begins to speak in colour phenomena. One feels how the nature spirits of earth and water rise up from oxalic acid and affirm themselves, but then how the whole passes over into a humming configuration of melodies, harmonies, which then return into themselves. Thus can be experienced this process that gives rise to formic acid and carbon dioxide.[81]

In the various pictures Steiner offers us in the lectures, he illustrates the diversity of the formic acid process. Below, these are presented in summary.

Within the earth organism, the same process occurs as in ourselves. There the carriers of the process are the insects, which are also the bearers of the astral life of the whole earth organism. In the previous chapter, this close connection between etheric and astral life was embodied in the image of the plant and animal clouds. Animal life creates the foundation for further plant life, and vice versa. The example of the carpenter bee shows that the insects draw their powers from breakdown, in this case from excavation and decomposition of wood. The carpenter bee can transform this plant material of dead wood into life again, thereby ensuring that plant matter does not lapse from the life process.

The death process in the plant world arises in autumn as nature dies away, but is countered in early spring again through the formic acid in the atmosphere. This instigates an awakening impulse, eliciting from seeds in the soil a desire to become plants once more. The formic acid process represents a continual renewal of life, and death arises where this ceases.

Rudolf Steiner regards the death process in a beehive in similar terms. If sufficient bee venom cannot be produced, with its high percentage of formic acid, the death process intervenes. Just as soul and spirit 'fly forth' when a person dies, so the swarm withdraws with its old queen. This death process is at the same time the precondition for the colony's new life. Every beekeeper can have an intuitive sense of this. The unity of the hive, the unity of the organism, is lost as the new queens develop. The nature of the colony changes. As the swarm flies off, enormous chaos arises—the air is full of bees. As soon as they form a cluster on a branch, a new colony life begins. The swarm can be caught with a light knock on the branch, and the bees falls unresisting and heavily into the box held ready to catch them. After a rest period in a cool cellar, this swarm will develop great vitality, and can build a great many combs in only a few weeks.

To see swarming as a process of death and renewal is highly illuminating: a swarm accomplishes a death process following which it comes to rest in the cluster. Its reincarnation in a new hive, with newly created combs, follows after this rest period. Rudolf Steiner shows here how life and death must be in reciprocal interplay in the whole of nature. For a colony to multiply without prior swarming does injury to this principle. Life forces are released specifically by allowing the old to pass away.

SOURCES

The following volumes are cited in this book. Where relevant, published editions of equivalent English translations are indicated. The works of Rudolf Steiner are listed with the volume numbers of the complete works in German, the *Gesamtausgabe* (GA), as published by Rudolf Steiner Verlag, Dornach, Switzerland.

RSP = Rudolf Steiner Press, UK
AP/SB = Anthroposophic Press/SteinerBooks, USA

GA

1 *Nature's Open Secret* (AP)
2 *Theory of Knowledge Implicit in Goethe's World Conception* (AP)
3 *Truth and Knowledge* (AP)
13 *Occult Science, An Outline* (RSP)
27 *Extending Practical Medicine* (RSP)
53 *Ursprung und Ziel des Menschen*
98 *Natur- und Geistwesen—ihr wirken in unserer sichtbaren Welt*
101 *Mythen und Sagen. Okkulte Zeichen und Symbole*
199 *Spiritual Science as Foundation for Social Forms* (AP)
211 *Sun Mystery and Mystery of Death and Resurrection* (SB)
230 *Harmony of the Creative Word* (RSP)
232 *Mystery Knowledge and Mystery Centres* (RSP)
233 *World History in the Light of Anthroposophy* (RSP)
314 *Physiology and Healing* (RSP)
316 *Understanding Healing* (RSP)
319 *The Healing Process* (SB)
347 *From Crystals to Crocodiles* (RSP)
348 *From Comets to Cocaine* (RSP)
351 *Bees* (SB)

All English-language titles are available via Rudolf Steiner Press, UK (www.rudolfsteinerpress.com) or SteinerBooks, USA (www.steinerbooks.org)

NOTES

Page references below for Steiner's works refer to volumes published in German

1. See Jürgen Tautz, *The Buzz About Bees*, Springer 2008; Jürgen Tautz, Helga R. Heilmann, *Phänomen Honigbiene*, Spektrum Akademischer Verlag, 2007.
2. Lecture in Dornach, 2 January 1924, GA 316, pp. 26f.
3. Lecture in Dornach, 3 February 1923, GA 348, pp. 316–18.
4. Remarks after a talk on bees by Mr Müller, Dornach, 10 November 1923, GA 351, pp. 131f.
5. Steiner often uses the word 'honey' when in fact 'nectar' is meant, and the translator has accordingly often used the word 'nectar' where this is the case.
6. Lecture in Dornach, 26 November 1923, GA 352, pp. 133–48.
7. Ibid., p. 146.
8. See: http://demeter.de/index.php?id=1521&MP=13-1491&no_cache=1&file=20&uid=343
9. Article by Professor Buttel-Reepen, Oldenburg, in, *Schweizerische Bienen-Zeitung N.F.*, no. 2, February 1923, pp. 85–7; and no. 3, March 1923, pp. 134–6.
10. The article (referred to above) by Buttel-Reepen in the *Swiss Beekeeper's Journal* cites experiments with ants by the Swiss psychiatrist and entomologist Auguste Forel (1848–1931), which are described in his book *Das Sinnesleben der Insekten* (Munich 1910). When pupae were uncovered in a completely darkened room, the insects only carried them into the interior of their nest if they [the pupae] were exposed to ultraviolet rays. Based on trials with ants whose eyes he painted with black pigment, he assumed that they saw ultraviolet light (ibid.,

p. 136). Training experiments by the zoologist Alfred Kühn (1885–1968) and the physicist Robert Wichard Pohl (1884–1976) drawing on experiments by the zoologist Karl von Frisch (1886–1982) are, in the view of Buttel-Reepen, 'so persuasive about bees' capacity to see colours (albeit limited in scope) that this ability can no longer be doubted'. Ibid., p. 135.

11. Honeydew collected from forest plants which the bees convert into forest honey.
12. Lecture in Dornach, 28 November 1923, GA 351, pp. 149–60.
13. See note 1.
14. See, among other works, 'Der Farben- und Formensinn der Bienen', in *Zoologische Jahrbücher (Physiologie) 35*, 1–188 (1914–15); and *Aus dem Leben der Bienen*, Berlin 1927 (10th edition 1993). English: *Bees: Their Vision, Chemical Senses and Language*, Cornell University Press, 1971.
15. GA 3, p. 90.
16. GA 1, p. 126.
17. GA 2, p. 101.
18. Ibid., p. 95.
19. Lecture in Berlin, 11 May 1905, GA 53, p. 435.
20. Lecture in Dornach, 26 November 1923, GA 351, p. 148.
21. In *Schweizerische Bienen-Zeitung N.F.*, no. 3, March 1923, pp. 136–42.
22. All case histories ibid., pp. 141f.
23. Ibid., p. 139. Gustav von Bunge (1844–1920) was a German-Baltic physiologist and from 1885 a professor at Basel University. His major works are: *Lehrbuch der physiologischen und pathologischen Chemie* (1887) and *Lehrbuch der Physiologie des Menschen* (1901).
24. The Polish-American biochemist Casimir Funk (1884–1967) coined the term 'vitamin' in 1912, composing it from the Latin *vita*, meaning life, and amines (organic bases, derived from ammonia).

25. This comment could not be located in the works of Heine (1797–1856). In Chapter 38 of *Ut Mine Stromtid* (Verlag der Hinstorff'schen Hofbuchhandlung, 1865, p. 134), which is the third novel of an autobiographical trilogy, *Olle Kamellen*, by Fritz Reuter (1810–74), Inspector Zacharias Bräsig concludes his talk at the Rahnstädt Reform Club with the following words: 'Fellow citizens, I wish to say this to you since I've been living here in the city long enough now and studying humanity: the great poverty in the city comes from penury!'
26. Paula Emrich, op. cit. (see note 21), p. 142.
27. Ibid., p. 136.
28. Ibid., p. 138.
29. Lecture in Dornach, 1 December 1923, GA 351, pp. 161–74.
30. Lecture in Dornach, 30 December 1923, GA 233, p. 127.
31. See note 1.
32. Lecture in Dornach, 2 January 1924, GA 314, p. 210.
33. GA 27, p. 76.
34. Lecture in London, 28 August 1924, GA 319, pp. 217f.
35. Wilhelm von Osten (1838–1909) was the son of an estate owner, and later became a teacher. He supposedly trained his horse, who became known as 'Clever Hans', to count and do sums. After von Osten's death, the Elberfeld jeweller Karl Krall (1863–1929) undertook experiments with this animal and others, and in 1912 published the book *Denkende Tiere* ('Thinking Animals'). See also a book by the German psychologist Oskar Pfungst (1874–1933), *Der Kluge Hans: Ein Beitrag zur nichtverbalen Kommunikation*, Frankfurt am Main 1983 (new edition of the original published in 1907 entitled *Das Pferd des Herrn von Osten 'Der kluge Hans': Ein Beitrag zur experimentellen Tier- und Menschen-Psychologie*).
36. See note 35 above.
37. Lecture in Dornach, 1 December 1923, GA 351, pp. 174–80.
38. Fr. Leuenberger: 'Honigreklame', in *Schweizerische Bienen-Zeitung N.F.*, no. 10, October 1923, pp. 471f.

39. Under the heading '*Rundschau*' ('Overview'), Fr. Leuenberger, 'Eine süsse Medizin', ibid. p. 474. This is a report from the *Bulletin de l'apiculture des Alpes maritimes* by the 'renowned beekeeper Ph. J. Baldensberger'.
40. 'An English chemist, Miss Elizabeth Sedney Semmers, recently ascertained that starch in plant cells is transformed into sugar through the influence of moonlight. The same author finds that seeds germinate quicker in moonlight. It is known that all sugar in plants is created through the transformation of starch. [...]' Under the heading 'Rundschau', 'Mondlicht und Honigproduktion', in *Schweizerische Bienen-Zeitung N.F.*, no. 11, November 1923, p. 532.
41. Honey from the honeydew of the white fir (*Abies alba*) which has a high fruit sugar and high ash content and can burden overwintering bees' digestion.
42. In July 1920, the physician and teacher Eugen Kolisko (1883–1939) worked with Rudolf Steiner on a remedy to combat foot-and-mouth disease, which was very serious at the time. The laboratory work necessary to discover the right dosage of the remedy was entrusted to scientist Lili Kolisko (1889–1976), who also published a paper on this: *Milzfunktion und Plättchenfrage*, Stuttgart 1922. See also Joseph Werr, *Tierzucht und Tiermedizin im Rahmen biologisch-dynamischer Landwirtschaft*, Stuttgart 1953.
43. The proposed lecture did not take place until Monday, 10 December 1923.
44. Lecture in Dornach, 5 December 1923, GA 351, pp. 181–95.
45. Numerous races of the European honey bee (*Apis mellifera*) developed, including the Italian bee of the Apennine peninsula (*Apis mellifera ligustica*), which has characteristic yellow rings on its rear abdomen, and the Slovenian or Carnica bee (*Apis mellifera carnica*), which originates in Carniola, the region around Ljubljana.
46. This was very probably the agriculturalist Hugo Hitschmann

(1838–1904), founder of the modern farming and forestry press in Austria.

47. Lecture in Dornach, 10 December 1923, GA 351, pp. 196–203.
48. See my obituary at http://www.summ-summ.ch/bibl/pub/h_wille.html
49. The gall wasps (*Cynipidae*).
50. Outer envelope of the embryo in human beings and mammals, by means of which substances are exchanged between mother and foetus.
51. The bedeguar gall wasp, *Rhoditis* (*Diplolepis*) *rosae*, makes galls with hairy excrescences which are given locally diverse names such as robin's pincushion, rose queen, rose sponge, etc.
52. Rudolf Steiner is clearly drawing here on *Brehms Illustriertes Tierleben*, home and school edition by Friedrich Schödler, Volume 3, Leipzig 1875, p. 511 (the relevant volume is in Steiner's library), where it is stated: 'It is well known that the ancients already made use of a gall wasp, *Cynips senes Linnaeus* (nowadays known as *Blastophaga psenes*) in order to obtain juicier and better tasting figs; and still today in Greece, great care is taken to "caprify" cultivated figs by using this insect. It lives in wild figs, and develops fully by the end of June before the figs have ripened. It would stay dwelling in them if it were not disturbed by human beings. But for this purpose people pick the figs, bind them in pairs with a long reed stalk and throw them over the branches of a cultivated fig, distributing them as evenly as possible between its fruits. As these wild figs dry out and shrink, it causes the insects to leave them, to emerge and start an (abnormal) second generation, choosing the cultivated figs as habitation for their brood. The figs are harvested before development of this second generation, which perishes, though not before it has increased the richness of sap in the host fruit.'
53. Lecture in Dornach, 10 December 1923, GA 351, pp. 203–12.

54. Lecture in Stuttgart, 14 September 1907, GA 101, pp. 167f.
55. This passage contradicts Steiner's earlier statement (see pages 18f.), according to which the *worker bee* corresponds to one revolution of the sun, which is there given as 21 days.
56. The term 'Tartarus', nowadays used primarily for gout and rheumatic conditions, can be traced back to Paracelsus, who understood it to mean metabolic diseases in general. In ancient times already, bee venom was used to treat rheumatism and arthritis.
57. Apis Mellifica.
58. Lecture in Dornach, 12 December 1923, GA 351, pp. 213–27.
59. Lecture in Dornach, 12 December 1923, GA 351, pp. 215f.
60. 'The agricultural ant (*Myrmica molificans* Dar.), in Texas protects its dwelling with a circular ridge up to 50 cm high, then clears and levels the ground surrounding this ridge for up to 1 metre, and here allows nothing to grow except for a particular type of grass, *Aristida oligantha*. It carefully cultivates this and harvests the ripe grains which are dehusked in one part of the nest and then stored away.' (*Meyers Konversationslexikon*, Leipzig and Vienna, fourth edition 1885–92, Vol. 1, p. 453.
61. See *Brehms Thierleben. Allgemeine Kunde des Thierreichs* (ninth volume, Part Four: *Wirbellose Tiere*, Volume 1: *Die Insekten, Tausendfüssler und Spinnen*, Leipzig 1884, p. 266): 'The agricultural ant (*Myrmica molificans*) was the subject of a talk which Darwin gave at the Linnean Society in London based on observations made by Linsecom in Texas.' This talk further details the behaviour of these ants.
62. The ichneumon or parasitic wasp (*Ichneumonidae*).
63. In a lecture in Dornach on 5 January 1923, in GA 348.
64. GA 13.
65. *Brehms Thierleben*, op. cit., p. 249: 'The ungoverned wildness of wasps will be sufficiently known to all, even if he has not—as happened once to me—been attacked by a whole swarm of

them and mercilessly stung simply because I was innocently walking along a path alongside which lay the entrance to their nest. A few years ago a sheepdog and his companions had a similar experience. There were cows grazing at a particular spot which contained a great many molehills. The dog sat down on one of these to faithfully guard the herd. All at once he let out a terrible wail and threw himself headlong into the stream flowing close by. The young cattle herder, not realizing what had happened, hurried to help his faithful animal, called to him and found him sprinkled all over with wasps. As he was engaged in removing the now somewhat cooled-off creatures from the dog, he did not notice that he himself was standing upon a volcano. The irritated wasps crawled up his legs, into his clothing, and higher, and he too had to seek refuge hurriedly in the water, suffering a great many stings in the process. The confusion increased: these molehills were inhabited by numerous colonies, which no one had known hitherto. The grazing cattle also had some encounters with the wasps, and were attacked by these now furious creatures. A great, general caterwaul of cries and alarms was the result, with many headlong dashes into the stream. It took much effort and the help of many to gradually restore order. Attempts to destroy these nests and to make the meadow available again for grazing remained fruitless. That year the wasps were too numerous and remained lords of the situation and the locality.'

66. Lecture in Dornach, 15 December 1923, ibid, pp. 228–44.
67. Ibid., p. 229.
68. Hans Wille, 'Überlebensstrategie des Bienenvolkes', *Bienenwelt*, Vol. 27, 1985, pp. 169–82.
69. Lecture in Dornach, 15 December 1923, op. cit., pp. 240f.
70. Ibid., p. 237.
71. Lecture in Dornach, 26 October 1923, GA 230, p. 73.
72. Lecture in Cologne, 7 June 1908, GA 98, pp. 88–90.
73. Jean-Henri Fabre, 1823–1915, a French entomologist. His

compendious work *Souvenirs Entomologiques* ('Entomological Memoirs') appeared in ten volumes in Paris between 1879 and 1907. In 2010, Matthes & Seitz, Berlin, began to publish this work in German.

74. *Xylocopa violacea* or *Xylocopa valga. Brehms Thierleben*, op. cit., pp. 226f: 'With loud buzzing the female flies, in pursuit of her brood duties, around poles, boarded walls or posts, lets the sun shine upon her and then buzzes off again. These movements are no doubt directed primarily to seeking out a suitable place where she can bring the next generation to birth, since her brief life belongs not to her but to her descendants. Old wood, a decaying post, a rotting tree trunk from which shreds of bark are missing, are the most suitable for her, and make her difficult work easier. Eagerly the bee gnaws a hole in the wood, of her own body's diameter, penetrates several millimetres into it then turns her attention downwards. For this work she requires a drill and forceps (which her two mandibles provide). She gets rid of the splinters and flakes of wood, and delves ever deeper until she has created a tube of even proportions, as long as 31 centimetres, whose end curves outwards again. The careful mother allows herself only as much rest during this work as is sufficient for excursions to nearby flowers, where she draws new strength from the nectar. In the lower storey of her delvings, she now mixes nectar with pollen in a specific quantity, then lays an egg upon this, and at a height roughly equal to the diameter of the tube she places a lid made of concentric rings of kneaded wood shavings. The first cell is thus closed, at the same time creating the floor for the next above it. Again, the same quantity of food is deposited, and again an egg is laid upon it. Thus she continues without interruption, so long as bad weather does not prevent her, until the available passage is filled with a column of such cells. [. . .] After a few days the [first] young grub hatches [. . .] This grub lies curled up, and by the time roughly three weeks have passed

it has grown to fill the cavity where it lies. A close examination shows black grains of excreta beside it. Now the grub spins itself a cocoon and pupates within. Since the oldest is lowermost in the column, naturally it must develop soonest, the second next and so on upwards to the last and latest developing pupa. Will she wait to emerge until the last of her sisters has opened the way out from her dungeon? In the case of the second brood, yes, since then winter prevents her from emerging. But this is not true of the first brood, which develops during August: the easiest method of escape becomes apparent. She is head downwards in the cavity and now need only move around a little and forward to find that the space opens before her. Coming to the end of the chamber, which is loosely filled with flakes of wood, she instinctively uses her mandibles to gnaw through the thin layer between her and the warm summer air. [. . .] The second to emerge simply follows the passage taken by the first, until finally the whole brood has flown and the nest is empty.'

75. Lecture in Dornach, 22 December 1923, GA 351, pp. 245–61.
76. Lecture in Dornach, 22 December 1923, GA 351, p. 257.
77. Lecture in Dornach, 3 September 1920, GA 199, p. 202.
78. Lecture in Dornach, 5 August 1922, GA 347, p. 34.
79. Lecture in Dornach, 2 April 1922, GA 211, p. 121.
80. Lecture in Dornach, 22 December 1923, GA 232, p. 195.
81. Ibid., p. 200.

Steiner

A NOTE FROM RUDOLF STEINER PRESS

We are an independent publisher and registered charity (non-profit organisation) dedicated to making available the work of Rudolf Steiner in English translation. We care a great deal about the content of our books and have hundreds of titles available – as printed books, ebooks and in audio formats.

As a publisher devoted to anthroposophy…

- We continually commission translations of previously unpublished works by Rudolf Steiner and invest in re-translating, editing and improving our editions.
- We are committed to making anthroposophy available to all by publishing introductory books as well as contemporary research.
- Our new print editions and ebooks are carefully checked and proofread for accuracy, and converted into all formats for all platforms.
- Our translations are officially authorised by Rudolf Steiner's estate in Dornach, Switzerland, to whom we pay royalties on sales, thus assisting their critical work.

So, look out for Rudolf Steiner Press as a mark of quality and support us today by buying our books, or contact us should you wish to sponsor specific titles or to support the charity with a gift or legacy.

office@rudolfsteinerpress.com
Join our e-mailing list at www.rudolfsteinerpress.com

RUDOLF STEINER PRESS